Rock Mechanics

AN INTRODUCTION

Rock Mechanics

AN INTRODUCTION

Nagaratnam Sivakugan
Sanjay Kumar Shukla
and Braja M. Das

CRC Press is an imprint of the
Taylor & Francis Group, an **informa** business

CRC Press
Taylor & Francis Group
6000 Broken Sound Parkway NW, Suite 300
Boca Raton, FL 33487-2742

Version Date: 20121026

International Standard Book Number: 978-0-415-80923-8 (Hardback)

Library of Congress Cataloging-in-Publication Data

Sivakugan, N. (Nagaratnam), 1956-
Rock mechanics : an introduction / N. Sivakugan, S.K. Shukla, Braja M. Das.
p. cm.
Includes bibliographical references and index.
ISBN 978-0-415-80923-8 (hardback)
1. Rock mechanics. I. Shukla, S. K. II. Das, Braja M., 1941- III. Title.

TA706.S49 2013
624.1'5132--dc23 2012032578

Visit the Taylor & Francis Web site at
http://www.taylorandfrancis.com

and the CRC Press Web site at
http://www.crcpress.com

Contents

Preface

Rock mechanics is a subject that is not commonly present in most undergraduate civil engineering curriculums worldwide. It is sometimes taught as an elective subject in the final year of the bachelor's degree program or as a postgraduate subject. Nevertheless, civil and mining engineers and academicians would agree on the usefulness and the value of some exposure to rock mechanics at the undergraduate level. Ideally speaking, engineering geology and rock mechanics are the two areas that should always be included in a comprehensive civil engineering curriculum. A good understanding of engineering geology and rock mechanics enables future practitioners to get a broader picture in many field situations. They are often the weakest links for many geotechnical/civil engineering professionals.

The main objective of this book is to present the fundamentals of rock mechanics with a geological base in their simplest form to civil engineering students who have no prior knowledge of these areas. There are also geological engineering degree programs that are offered in many universities that would find the book attractive.

This book is authored by three academicians who have written several books in geotechnical engineering and related areas and have proven track records in successful teaching. We thank all those who have assisted in preparing the manuscripts and reviewing the drafts, as well as all those who provided constructive feedback. The support from Simon Bates of the Taylor & Francis Group during the last two years is gratefully acknowledged.

Nagaratnam Sivakugan, Sanjay Kumar Shukla and Braja M. Das

Authors

Dr. Nagaratnam Sivakugan is an associate professor and head of Civil and Environmental Engineering at James Cook University, Townsville, Australia. He graduated with a first-class honours degree from the University of Peradeniya – Sri Lanka and received his MSCE and PhD from Purdue University, USA. He is a fellow of Engineers Australia, a chartered professional engineer and a registered professional engineer of Queensland. He does substantial consulting work for geotechnical and mining companies in Australia and for some overseas organisations including the World Bank. He serves on the editorial boards of the *International Journal of Geotechnical Engineering* and *Indian Geotechnical Journal*. He is the coauthor of 3 books, 7 book chapters, 90 refereed international journal papers and 60 international conference papers.

Dr. Sanjay Kumar Shukla received his BSc in civil engineering (1988) from Ranchi University, Ranchi, India; MTech in civil engineering (1992) from Indian Institute of Technology Kanpur, Kanpur, India and PhD in civil engineering (1995) from Indian Institute of Technology Kanpur, Kanpur, India. He is an associate professor and program leader of the Discipline of Civil Engineering at the School of Engineering, Edith Cowan University, Australia. He has more than 20 years of teaching, research and consultancy experience in the field of geotechnical and geosynthetic engineering. He has authored 115 research papers and technical articles including 72 refereed journal publications. Currently on the editorial board of the *International Journal of Geotechnical Engineering*, USA, Sanjay is a fellow of the Institution of Engineers Australia, a life fellow of the Institution of Engineers (India) and the Indian Geotechnical Society.

Dr. Braja M. Das received his BSc degree with honors in physics (1959) from Utkal University, Orissa, India; BSc in civil engineering (1963) from Utkal University, Orissa, India; MS in civil engineering (1967) from University of Iowa, USA and PhD in geotechnical engineering (1972) from University of Wisconsin, USA. He is the author of several geotechnical engineering texts and reference books. A number of these books have been translated into several languages and are used worldwide. He has authored more than 250 technical papers in the area of geotechnical engineering. He is a fellow and life member of the American Society of Civil Engineers as well as an emeritus member of the Committee on Chemical and Mechanical Stabilization of the Transportation Research Board of the National Research Council of the United States. From 1994 to 2006, he served as Dean of the College of Engineering and Computer Science at California State University, Sacramento.

Chapter 1

Fundamentals of engineering geology

1.1 INTRODUCTION

The earth materials that constitute relatively the thin outer shell, called *crust*, of the Earth are arbitrarily categorised by civil engineers as *soils* and *rocks*. These materials are made up of small crystalline units known as *minerals*. A mineral is basically a naturally occurring inorganic substance composed of one or more elements with a unique chemical composition, unique arrangement of elements (crystalline structure) and distinctive physical properties.

Soils and *rocks* have various meanings among different disciplines. In civil engineering, the *soil* is considered as a natural aggregate of mineral grains that can be separated by gentle mechanical means such as agitation in water. It comprises all the materials in the surface layer of the Earth's crust that are loose enough to be normally excavated by manual methods using spade or shovel. The *rock* is a hard, compact and naturally occurring earth material composed of one or more minerals and is permanent and durable for engineering applications. Rocks generally require blasting and machinery for their excavation. It should be noted that geologists consider engineering soils as unconsolidated rock materials composed of one or more minerals. One rock is distinguished from the other essentially on the basis of its mineralogical composition.

Geology is the science concerned with the study of the history of the Earth, the rocks of which it is composed and the changes that it has undergone or is undergoing. In short, geology is the science of rocks and earth processes. *Engineering geology* deals with the application of geologic fundamentals to engineering practice. *Rock mechanics* is the subject concerned with the study of the response of rock to an applied disturbance caused by natural or engineering processes. *Rock engineering* deals with the engineering applications of the basic principles and the information available in the subjects of engineering geology and rock mechanics in an economic way. All these subjects are closely concerned with several engineering disciplines such as civil, mining, petroleum and geological engineering.

Rock mechanics is a relatively young discipline that emerged in the 1950s, two decades after its sister discipline of soil mechanics. The failure of Malpasset concrete arch dam in France (Figure 1.1a) on December 3, 1959, killing 450 people, and an upstream landslide that displaced a large volume of water, overtopping Vajont Dam in Italy (Figure 1.1b) on October 9, 1963, claiming more than 2000 lives downstream, were two major disasters that triggered the need for better understanding and more research into rock mechanics principles. The first proper rock mechanics textbook *La Mécanique des Roches* was written by J.A. Talobre in 1957. Rock mechanics is a multidisciplinary subject relating geology, geophysics and engineering, which is quite relevant to many areas of civil, mining, petroleum and geological engineering. Good grasp of rock mechanics would be invaluable to civil engineers, especially to those who specialise as geotechnical engineers. Here, we apply the principles we learned in mechanics to study the engineering behaviour of the rock mass

(a)

(b)

Figure 1.1 Dam failures: (a) Malpasset after failure and (b) Vajont dam currently.

in the field. Rock mechanics applications include stability of rock slopes, rock bolting, foundations on rocks, tunnelling, blasting, open pit and underground mining, mine subsidence, dams, bridges and highways.

This chapter presents the geological fundamentals with their relations to engineering. These concepts are required to understand rock mechanics and its applications in a better way.

1.2 STRUCTURE AND COMPOSITION OF THE EARTH

The shape of the Earth is commonly described as a spheroid. It has an equatorial diameter of 12,757.776 km and a polar diameter of 12,713.824 km. The total mass of the Earth is estimated as 5.975×10^{24} kg and its mean density as 5520 kg/m^3. Detailed scientific studies have indicated that the Earth is composed of three well-defined shells: *crust, mantle* and *core* (Figure 1.2). The topmost shell of the Earth is the *crust*, which has a thickness of 30–35 km in continents and 5–6 km in oceans. The oceanic crust is made up of heavier and darker rocks called basalts while the continental crust consists of light-coloured and light-density granitic rocks. The Earth is basically an elastic solid, and when expressed in terms of oxides, it has silica (SiO_2) as the most dominant component, its value lying more than 50% by volume in oceanic crust and more than 62% in the continental crust. Alumina (Al_2O_3) is the next important oxide varying between 13% and 16%. The zone of materials lying between the crust and a depth of 2900 km is known as the *mantle*, which is made up of extremely *basic* materials (very rich in iron and magnesium but quite poor in silica). The mantle is believed to be highly plastic or ductile solid in nature. The innermost structural shell of the Earth known as the *core* starts

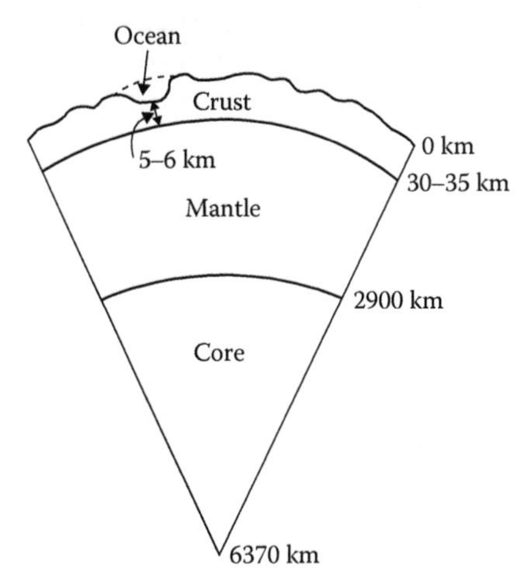

Figure 1.2 Structure of the Earth (*Note*: not to scale).

at a depth of 2900 km below the surface and extends right up to the centre of the Earth at 6370 km. The materials of the core are probably iron and nickel alloys. The outer core is believed to have no shear resistance, which makes it almost a liquid, whereas the inner core is a ductile solid. The core has a very high density, more than 10,000 kg/m^3, at the mantle–core boundary.

Lithosphere (Greek: *lithos* = stone) is a combination of the Earth's crust and the outer part of the upper mantle. It is an elastic solid. Its thickness is approximately 100 km. *Asthenosphere* is the upper mantle, which is ductile and 3% liquid (partially melting). Its thickness is approximately 600 km.

Below the Earth's surface, the temperature increases downwards at an average rate of 30°C/km. This rate is higher near a source of heat such as an active volcanic centre and is also affected by the thermal conductivity of the rocks at a particular locality. Based on this rate, a simple calculation shows that at a depth of around 30–35 km, the temperature would be such that most of the rocks would begin to melt. The high pressure prevailing at that depth and the ability of crustal rocks to conduct heat away to the surface of the Earth help the rock material there to remain in a relatively solid condition, but there will be a depth at which it becomes essentially a viscous fluid and this defines the base of the lithosphere.

1.3 MINERALS AND MINERALOGICAL ANALYSIS

Minerals are the building blocks for soils and rocks present in the Earth, and they have distinctive physical properties, namely *colour, streak, hardness, cleavage, fracture, lustre, habit (or form), tenacity, specific gravity, magnetism, odour, taste* and *feel*. The streak of a mineral is the colour of its powder. The hardness of a mineral is its resistance to abrasion. The cleavage of a mineral is its tendency to break down along a particular direction; it is described as one set of cleavage, two sets of cleavage and so on. Fracture is the character of the broken surface of the mineral in a direction other than the cleavage direction. Lustre is the appearance of the mineral in reflected light. Habit (or form) of a mineral refers to the size and shape of its crystals. Tenacity describes the response of a mineral to hammer blows, to cutting with a knife and to bending.

Hardness and specific gravity are the most useful diagnostic physical properties of a mineral. Hardness is tested by scratching the minerals of known hardness with a specimen of the mineral of unknown hardness. In practice, a standard scale of 10 minerals, known as the Mohs scale of hardness (see Table 1.1), is used for this purpose. The hardness of minerals listed in Table 1.1 increases from 1 for talc to 10 for diamond.

The specific gravity of a mineral is the ratio of its weight to the weight of an equal volume of water at a standard temperature, generally 4°C. The specific gravity of the common silicate minerals forming soils and rocks is about 2.65. For minerals forming the ores, the specific gravity may be as

Table 1.1 Mohs scale of hardness

Hardness	*Mineral*
1	Talc
2	Gypsum
3	Calcite
4	Fluorite
5	Apatite
6	Orthoclase
7	Quartz
8	Topaz
9	Corundum
10	Diamond

Table 1.2 Specific gravity of some common minerals

Mineral	*Specific gravity*
Apatite	3.2
Calcite	2.71
Chlorite	2.6–3.3
Clay minerals	2.5–2.8
Dolomite	2.85
Feldspar	2.56–2.7
Garnet	3.7–4.3
Gypsum	2.32
Hornblende	3.2
Halite	2.16
Hematite	4.72
Magnetite	5.2
Pyrite	5.01
Muscovite	2.8–3.0
Quartz	2.65
Rutile	4.2
Topaz	3.6
Tourmaline	3.0–3.2
Zircon	4.7

high as 20, for example, native platinum has a specific gravity of 21.46. Most minerals have specific gravity in a range of 2–6. Table 1.2 provides the specific gravity values of some common minerals.

Minerals are basically naturally occurring inorganic substances; however coal and petroleum, though of organic origin, are also included in the list of minerals. Almost all minerals are solids; the only exceptions are mercury, water and mineral oil (oil petroleum).

Table 1.3 Essential rock-forming minerals

Silicates	*Carbonates*
Silica (SiO_2)	Calcite (Ca carbonates)
Feldspars (Na, K, Ca and Al silicates)	Dolomite (Ca–Mg carbonates)
Amphiboles (Na, Ca, Mg, Fe and Al silicates)	
Pyroxenes (Mg, Fe, Ca and Al silicates)	
Micas (K, Mg, Fe and Al silicates)	
Garnets (Fe, Mg, Mn, Ca and Al silicates)	
Olivines (Mg and Fe silicates)	
Clay minerals (K, Fe, Mg and Al silicates)	

In civil engineering practice, it is important to have knowledge of the minerals that form the rocks; such minerals are called *rock-forming minerals*. Silicates and carbonates, as listed in Table 1.3, are the essential rock-forming minerals. Silicate minerals form the bulk (about 95%) of the Earth's crust. Silica and feldspars are the most common silicate minerals in the crust. Silica is found in several crystalline forms such as *quartz, chalcedony, flint, opal* and *chert*; quartz is one of the most common forms of silica. High quartz content in a rock indicates that it will have high strength and hardness. Feldspars form a large group of minerals; *orthoclase* or *K-feldspar* ($KAlSi_3O_8$), *albite* ($NaAlSi_3O_8$) and *anorthite* ($CaAlSi_2O_8$) are the main members. The mixtures (solid solutions) of albite and anorthite in different proportions form a series of feldspars called *plagioclases*. A plagioclase containing 40% albite and 60% anorthite is called *labradorite* and denoted as $Ab_{40}An_{40}$. K-feldspars alter readily into *kaolinite*, which is one of the clay minerals. *Hornblende* is a major mineral of the amphibole group of minerals. *Enstatite* ($MgSiO_3$), *hypersthene* [$(MgFe)SiO_3$] and *augite* [$(CaMgFeAl)_2(SiAl)_2O_6$] are the major minerals of the pyroxene group of minerals. There are two common types of micas: *muscovite* (*white mica*) [$KAl_2(Si_3Al)O_{10}(OH)_2$], which is rich in aluminium and generally colourless, and *biotite* (*black mica*) [$K(MgFe)_3(Si_3Al)O_{10}(OH)_2$], which is rich in iron and magnesium and generally dark brown to nearly black. Both types occur in foliated form and they can be split easily into thin sheets. The composition of common olivine is [$(MgFe)_2SiO_4$]. Since olivine crystallises at a high temperature (higher than 1000°C), it is one of the first minerals to form from the molten rock material called *magma*. Garnets occur both as essential and as accessory minerals in rocks. Clay minerals are hydrous aluminium silicates. *Kaolinite* [$Al_4Si_4O_{10}(OH)_8$], *illite* [$K_xAl_4(Si_{8-x}Al_x)O_{20}(OH)_4$, x varying between 1 and 1.5] and *montmorillonite* [$Al_4Si_8O_{20}(OH)_4$] are the principal clay minerals, which are described in detail in Section 1.6. *Calcite* ($CaCO_3$) and *dolomite* [$CaMg(CO_3)_2$] are carbonate minerals present in some rocks.

In addition to essential minerals, there are *accessory minerals* such as *zircon, andalusite, sphene* and *tourmaline*, which are present in relatively small proportions in rocks. Some minerals such as *chlorite, serpentine, talc, kaolinite* and *zeolite* result from the alteration of pre-existent minerals, and they are called *secondary minerals*. Since these minerals have little mechanical strength, their presence on joint planes within the jointed rock mass can significantly reduce its stability.

The common rock-forming minerals can be identified in the hand specimen with a magnifying glass, especially when at least one dimension of the mineral grain is greater than about 1 mm. With practice, much smaller grains can also be identified. This task is easily done by experienced geologists. If it is difficult to identify minerals by physical observations and investigations, X-ray diffraction and electron microscopic analyses make the identification task easy. Figure 1.3 shows a typical X-ray diffractogram of an air-dried clay fraction (<2 μm) collected from a shear surface of a recent landslip in South Cotswolds, United Kingdom, where clay minerals (kaolinite, K; illite, I and montmorillonite, M) are easily identified on the basis of a series of peaks of different intensities of X-rays reflected from the minerals corresponding to different angular rotations (2θ) of the detector of the X-ray diffractometer.

Figure 1.4 shows the photographs of some typical rock-forming minerals.

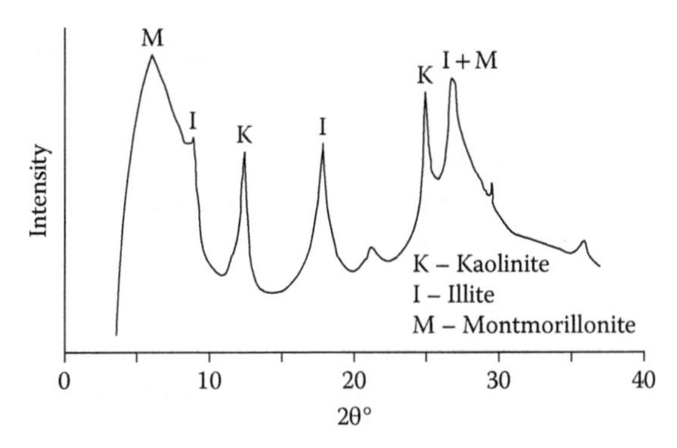

Figure 1.3 X-ray diffractogram of air-dried clay fraction (<2 μm). (Adapted from Anson, R.W.W. and A.B. Hawkins, *Geotechnique*, 49, 33–41, 1999.)

Figure 1.4 Photographs of some typical rock-forming minerals: (a) quartz, (b) orthoclase, (c) plagioclase, (d) muscovite, (e) biotite, (f) andradite garnet, (g) calcite, (h) dolomite and (i) chlorite. (Courtesy of Sanjay Kumar Shukla.)

Figure 1.4 (Continued)

1.4 ROCK FORMATIONS AND TYPES

Rocks form a major part of the Earth's crust. They are formed by the following processes:

1. Cooling of molten material (*magma*)
2. Settling, depositional or precipitation processes
3. Heating or squeezing processes

These three processes form the basis for rock classification and are also significant factors in establishing the mechanical properties of rocks. On the basis of their formation, rocks are classified as follows:

1. Igneous rocks
2. Sedimentary rocks
3. Metamorphic rocks

Rocks derived from magma are called *igneous rocks*, which are usually hard and crystalline in character. Igneous rocks make up about 95% of the volume of the Earth's crust. Some examples are *granite, basalt, dolerite, gabbro, syenite, rhyolite* and *andesite*. The silicates are the common igneous rock-forming minerals. There are six of them: silica, feldspars, amphiboles, pyroxenes, micas and olivine. Granite is usually light coloured (white, reddish, greyish etc.) and has a medium specific gravity, feldspar and quartz are the essential minerals and grains are medium or coarse. Rhyolite is mostly light coloured (light grey, yellow, pale red etc.) and has low specific gravity; grains are extremely fine and therefore constituent minerals cannot be easily identified. Basalt is dark grey or black in colour and has high specific gravity; mineral grains are too fine to be identified.

Igneous rocks are also known as *primary rocks* since these were the first formed rocks on the surface of the Earth. The characteristics of the igneous rocks are controlled by two basic factors: the rate of cooling when they were formed and the chemical composition of the magma. Rapid cooling precludes the growth of crystals, while slow cooling allows their growth. The igneous rocks produced due to rapid cooling of magma upon the surface of the Earth are known as *extrusive igneous rocks*, whereas those formed underneath the surface of the Earth due to slow cooling are known as *intrusive igneous rocks*. For example, basalt, rhyolite and andesite are extrusive igneous rocks, whereas granite, dolerite, gabbro and syenite are intrusive igneous rocks.

On the basis of silica content, igneous rocks are broadly classified as (1) acidic (>66% of silica), (2) intermediate (between 55% and 66% of silica), (3) basic (between 44% and 55% of silica) and (4) ultrabasic (<44% of silica) (Mukerjee, 1984). Granite, rhyolite and pegmatite are acidic igneous rocks, whereas basalt, dolerite and gabbro are basic igneous rocks.

Field observations of igneous rocks are very important for the determination of structure and extent of exposed rock mass. Geological maps and satellite imagery are useful for the determination of mode of occurrence of rocks in the field. In civil engineering constructions, particularly for large structures, the extent and occurrence of igneous rocks must be known.

The products of weathering (disintegration of rocks, see Section 1.6) are subjected, under favourable conditions, to transportation mostly by natural agencies such as running water, wind, glaciers and gravity, deposition and subsequent compaction or consolidation, resulting in *sedimentary rocks*. Some examples are *sandstone, shale, conglomerate, breccias, limestone, coal* and *evaporites*. Minerals forming the sedimentary rocks are kaolinite, illite, smectite, hematite, rutile, corundum, calcite, dolomite, gypsum, halite and so on. Sandstone is available in variable colours, and shades of grey, yellow, brown and red are frequent; it has low to medium specific gravity, and grains are rounded or angular and are cemented together by siliceous, calcareous or ferruginous material. Sandstone is usually massive, but bedded structure may sometimes be visible. Limestone is generally fine-grained and is found in lighter shades; calcite is the main constituent, although clay minerals, quartz, dolomite and so on may also be present. Conglomerate has different shades of colour, and the fragments are generally rounded.

Rocks that have undergone some chemical or physical changes subsequent to their original form are called *metamorphic rocks*. The process by which the original character or form of rocks is more or less completely altered is called *metamorphism*. This is mainly due to four factors: temperature, uniform pressure, directed pressure and access to chemically reactive fluids. Metamorphism brings changes in mineral composition and changes in texture of rock. Examples are *quartzite, slate, mica schist, marble, graphite, gneiss* and *anthracite.* Common metamorphic minerals are serpentine, talc, chlorite, kyanite, biotite, hornblende, garnet and so on. Quartzite, formed from sandstone with high silica content, is light coloured with shades of grey, yellow, pink and so on and has medium specific gravity. Slate, formed from shale, is a black or brown rock with low or medium specific gravity. Marble, formed from limestone, is commonly light coloured (white, grey, yellow, green, red etc.) and has a medium specific gravity; calcite is the main constituent of marble and dolomite is frequently associated with it.

In nature, one type of rock changes slowly to another type, forming a rock cycle (Figure 1.5). At the surface of the Earth, igneous rocks are exposed to weathering resulting in sediments, which may become sedimentary rocks due to hardening or cementation. If sedimentary and metamorphic rocks are deeply buried, the temperature and pressure may turn them into metamorphic rocks. Intense heat at great depths melts metamorphic and sedimentary rocks and produces magma, which may rise up and reach the Earth's surface where it cools to form igneous rocks.

Figure 1.6 shows photographs of some common types of rocks.

All kinds of rocks in the form of dressed blocks or slabs, called *building stones*, or in any other form, called *building materials*, are frequently used in civil engineering projects. Building stones are used in the construction of buildings, bridges, pavements, retaining walls, dams, docks and harbours and other masonry structures. Building materials are used as fine

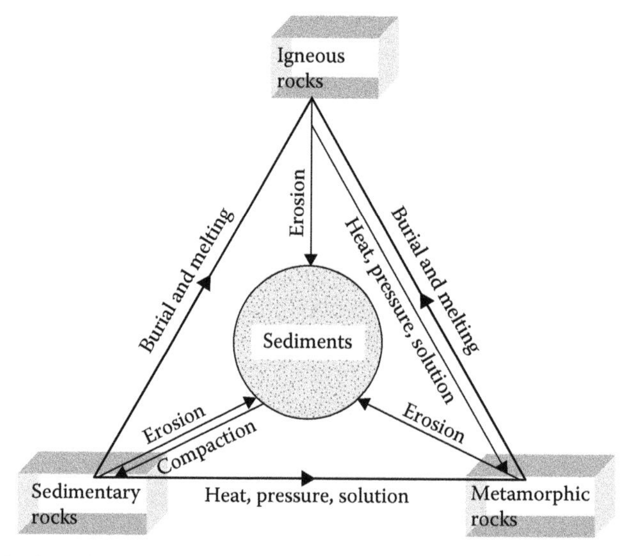

Figure 1.5 Rock cycle. (Adapted from Raymahashay, B.C., *Geochemistry for Hydrologists*, Allied Publishers Ltd., New Delhi, 1996.)

Figure 1.6 Photographs of some typical rocks: (a) granite, (b) basalt, (c) rhyolite, (d) sandstone, (e) limestone, (f) conglomerate, (g) marble, (h) slate and (i) mica schist. (Courtesy of Sanjay Kumar Shukla.)

and coarse aggregates in cement and bituminous concrete, raw materials in the manufacture of lime and cement, soils in making embankments and dams, ballasts in railway tracks, aggregates in subbase and base courses of highway and runway pavements and so on. As building stones and materials, rocks should have high strength and durability, which depend on their mineralogical composition, texture and structure. If the minerals of rocks are hard, free from cleavage and resistant to weathering, when these rocks are used as building stones and materials, they are likely to be strong and durable. The rock granite, composed mainly of quartz and feldspar, is very strong and durable, while carbonate rocks like marble and limestone are relatively weak and are worn out more rapidly. The rock quartzite, composed mainly of quartz alone, is obviously strong and durable, while mica schist is rather weak since it contains a lot of mica, which is an easily cleavable material. In crystalline rocks of igneous and metamorphic origin, the mineral grains are mutually interlocked and no open space is usually left in between the constituent grains. The interlocking texture of the mineral grains contributes substantially towards the strength of the crystalline rocks and the impervious nature of these rocks enables them to resist weathering.

In most sedimentary rocks, the mineral grains are held together by cementing materials of variable strength, and such rocks are generally porous due to the presence of voids/open spaces in them. The nature of the cementing materials determines the strength and durability of these rocks. Compared to igneous and metamorphic rocks, sedimentary rocks are weaker and less durable. Granites, marbles and gneisses are, thus, stronger and more durable than sandstones, limestone and conglomerates. Experience has shown that granites, gneisses and fine-grained and well-cemented sandstones last for centuries while limestone and coarse-grained and poorly cemented sandstones generally have a much shorter lifespan.

For the selection of rocks as building stones and materials, their mineralogical composition and texture should be studied carefully, and at the same time, their structural features should be closely observed in field. It is also necessary to determine their porosity and absorption; crushing and flexural strength; resistance to frost, fire and abrasion; modulus of elasticity and other properties of interest in the specific field applications.

1.5 GEOLOGICAL STRUCTURES AND DISCONTINUITIES

Geological structures, such as *folds, faults, joints* and *unconformities*, encountered in geology are regularly encountered in civil engineering work. For describing these structures, it is essential to understand the geometrical concept of the orientation of a plane and a line in space as described in detail in Section 2.2.

Orientation (or *attitude*) of a plane (rock bed, discontinuity plane or sloping ground) in space is described in terms of *strike* (S-S) and *dip* (ψ), or *dip* (ψ) and *dip direction* (D) (Figure 1.7). The *strike* of a plane is the direction of a line considered to be drawn along the plane so that it is horizontal. It is basically a trace of the intersection of the inclined plane with

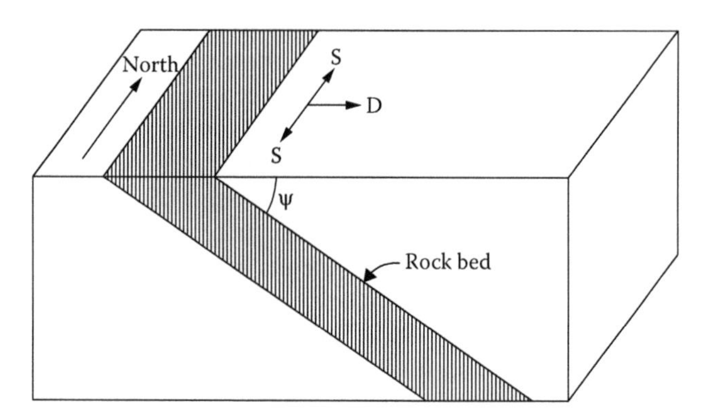

Figure 1.7 Dip and strike.

the horizontal reference plane. It is obvious that there will be only one such direction for any particular rock bed/discontinuity plane/sloping ground. The line of maximum inclination on the inclined plane is called the *line of dip*. The *dip* (ψ) of a plane is its maximum inclination to the horizontal plane, measured at right angles to the strike. For a horizontal plane, the dip is 0° and for a vertical plane, the same is 90°. Dip always refers to the *true dip*. *Apparent dip* is the inclination of any arbitrary line on the plane to the horizontal, and it is always smaller than the true dip. *Dip direction* (or *dip azimuth*) is the direction of the horizontal trace of the line of dip, expressed as an angle (α) measured clockwise from the north. It varies from 0° to 360°. In Figure 1.7, the rock bed strikes north–south, and therefore α is 90°.

Folds are defined as wavy undulations developed in the rocks of the Earth's crust due to horizontal compression resulting from gradual cooling of the Earth's crust, lateral deflection and intrusion of magma in the upper strata. Figure 1.8 shows a typical fold at an excavated site. Different elements of a fold are shown in Figure 1.9.

An *anticline* is an upfold where the limbs dip away from the axis of fold on either side. A *syncline* is a downfold where the limbs dip towards the axis of the fold on either side. Anticline and syncline can be noticed easily in Figure 1.8. The highest point on the arch of an anticline is called the *crest* of the fold and the lowest point on the syncline is called the *trough*. The sloping sides of a fold are called *limbs*. A reference plane that divides a fold into two equal halves is called an *axial plane*. The line of intersection of the axial plane and the surface of any constituent rock bed is called the *axis of the fold*, the inclination of which with the horizontal is called the *plunge*

Figure 1.8 Folded rock beds. (Courtesy of Dr. Dajkumar S. Jeyaraj.)

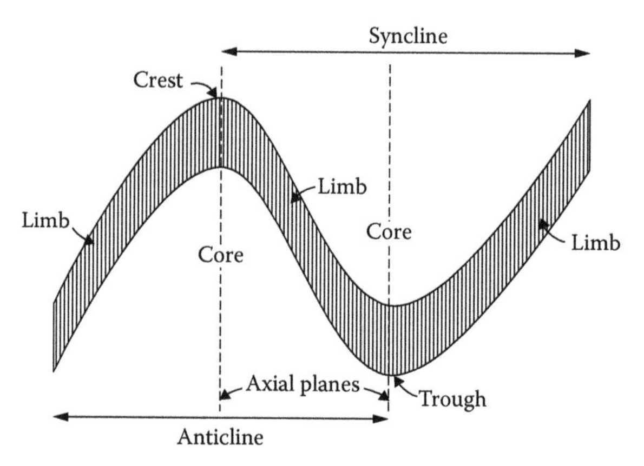

Figure 1.9 Elements of folds (anticline and syncline).

of the fold. In anticlines, the older rock beds generally occupy a position in the interior (or core) of the curvature, whereas in synclines, the rock beds in the interior are generally the youngest beds (Figure 1.9).

Faults are fractures in crustal strata along which appreciable shear displacement of the adjacent rock blocks have occurred relative to each other, probably due to tectonic activities. The fracture along which the shear displacement has taken place is called a *fault plane*. In general, the term 'fault' includes both the fault plane and the displacement that has occurred along it.

Figure 1.10 shows inclined faults, namely *normal fault* (Figure 1.10a) and *reverse fault* (Figure 1.10b), where the fault plane is inclined to the vertical. The total displacement AC that occurs along the fault plane is called the *net slip*. The amount of net slip may vary from only a few tens of millimetres to several hundred kilometres. The vertical component AB of the net slip AC is called the *throw* or *vertical slip*, and the horizontal component BC of the net slip AC is called the *heave* or *horizontal slip*. The angle subtended between the fault plane and any vertical plane striking in the same direction is called the *hade* of the fault. In Figure 1.10, $\angle BAC$ is the hade of the fault. It is observed that the two blocks lying on either side of the inclined fault plane are dissimilar in their configuration and orientation in space. Of these two adjacent blocks, one appears to rest on the other. The former is known as the hanging-wall (HW), while the latter, which supports the HW, is called the foot-wall (FW). In the normal fault, the HW appears to have moved relatively downwards in comparison with the adjoining FW, the whereas in the reverse fault, the FW appears to have been shifted downwards in comparison with the adjoining HW. From the mechanics point of view, the presence of tensile stresses causes the development of normal faults while compressive stresses lead to the formation of reverse faults. Fault plane, net slip, throw, heave and hade are called the *elements of the fault*.

Discontinuity is a collective term used for all structural breaks (bedding planes, fractures and joints) in solid geologic materials that usually have zero to low tensile strength. *Bedding planes* occur in sedimentary rocks due to disruption of the sedimentation process or repeated sedimentation cycles where the material deposited varies between cycles, generally on a geological time scale, which is defined in units of one million years, considering the estimated age of 4600 million years for the Earth. A *fracture* is where the continuity of the rock mass breaks. A *joint* is a fracture where little or no movement has taken place. This is the most common form of discontinuity encountered. These discontinuities can occur in several sets and are approximately parallel within a specific set (Figure 1.11). A *discontinuity set* is

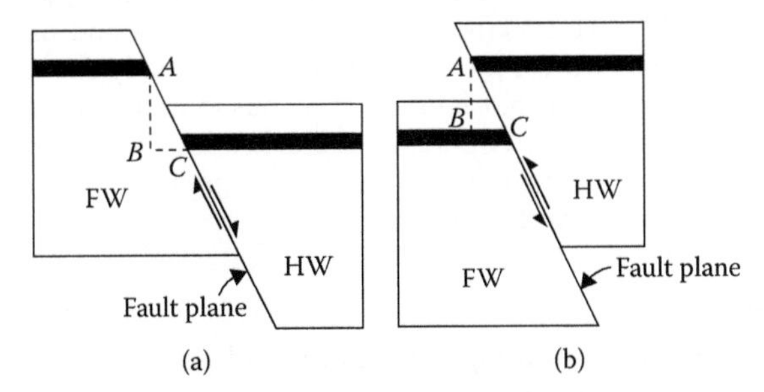

Figure 1.10 Inclined faults: (a) normal fault and (b) reverse fault.

Figure 1.11 Heavily jointed rock mass at 0.080 km of the Meja-Jirgo link canal, the site for the construction of a canal head regulator, Mirzapur, Uttar Pradesh, India. (After Shukla, S.K., Allowable Bearing Pressure for the Foundation Rock/Soil at km 0.080 of the Meja-Jirgo Link Canal for the Proposed Construction of a Hydraulic Structure (Head Regulator), Mirzapur, UP, India. A technical report dated 11 June 2007, Department of Civil Engineering, Institute of Technology, Banaras Hindu University, Varanasi, India, 2007.)

a series of discontinuities that have the same geologic origin, orientation, spacing and mechanical characteristics. The discontinuities make the rock mass anisotropic. More details about the rock mass and discontinuities are given in Chapter 4.

EXAMPLE 1.1

In field situations, the fault plane can be vertical, and the fault is known as the *vertical fault*. Do you think the terms *hanging-wall* and *foot-wall* are applicable here?

Solution

In the case of a vertical fault, the rock blocks on either side of the fault-plane will have exactly the same configuration and orientation in space; in other words, the structure remains exactly the same irrespective of whether one block or the other has moved relatively downwards. Therefore, of the two adjacent blocks, one does not appear to rest on the other, and therefore, the terms *hanging-wall* and *foot-wall* are not applicable to vertical faults.

EXAMPLE 1.2

Can you write a relationship between the hade and the dip of a fault plane?

Solution

In Figure 1.11, $\angle BAC$ is the hade of the fault plane, and $\angle ACB$ is the dip of the fault plane. Since $\angle ABC$ is a right-angled triangle, $\angle BAC + \angle ACB = 90°$.

The *plane of unconformity* or simply *the unconformity* is the surface/plane of separation between two series of rock beds/geological formations that belong to two different geologic ages and they are, in most cases, different in their geologic structure. The intersection of the plane of unconformity with the ground surface/topography constitutes the *line of unconformity* in the geological map. A geological map of an area exhibits the outcrops (portions of rocks exposed on the surface of the Earth) of the different rock types and geological formations and structures of that area, superimposed upon its topographical map. On a geological map, dashed lines represent the contour lines (imaginary lines that connect points of equal elevations) with the help of which the topographic features of the area are shown, and continuous lines represent the boundaries between the outcrops of rock beds (Figure 1.12).

When two series of beds are mutually related unconformably, the following relations generally hold good: (1) some of the beds appear to cover up some other beds; (2) the boundary of the latter appear to end abruptly against that of the former; (3) the dips of the beds differ in the two different formations thus related and (4) a conglomerate bed is often located above an unconformity. The unconformity signifies a time gap between the deposition/formation of one series of beds and the other. It is either a surface of erosion or non-deposition. In the field, an unconformity is commonly evidenced by the presence of a conglomerate bed. In Figure 1.12, the bed *A* is horizontal. The beds *B*, *C*, *D* and *E* slope towards the west. The boundary of the bed *A*, marked by the line *x-x*, is the line of unconformity in the map. The bed *A* is younger than the other group of beds (*B*, *C*, *D* and *E*) and appears to cover them up.

A careful study of geological structures and orientation of rock beds is essential for selecting the most suitable sites for civil engineering structures, and it also helps in planning safe excavations of open pits, shafts, stopes and tunnels in civil and mining projects. For example, a site with horizontal rock beds is the most capable of supporting the weight of building structures, but such sites may not be an ideal site for dams where water in the reservoir applies a horizontal force (thrust) on the dam embankment and sufficient seepage of water is expected, resulting in loss of reservoir water. Rock beds dipping upstream in the foundation may be the most competent to support the combined load *R* due to the weight (*W*) of the dam and thrust (*T*) from water in the reservoir (Figure 1.13). Additionally, such dipping rock beds do

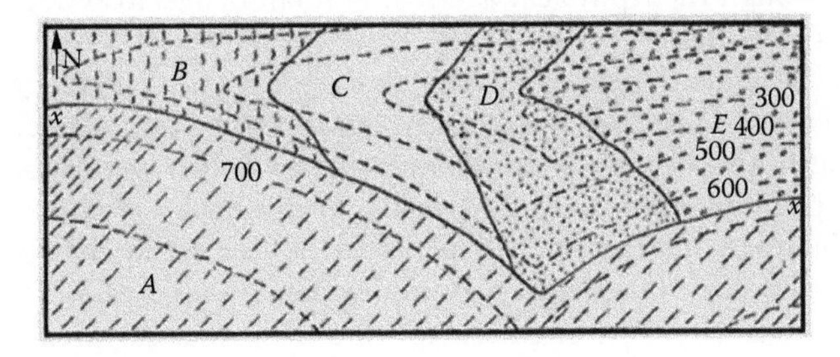

Figure 1.12 A typical geological map with the presence of an unconformity.

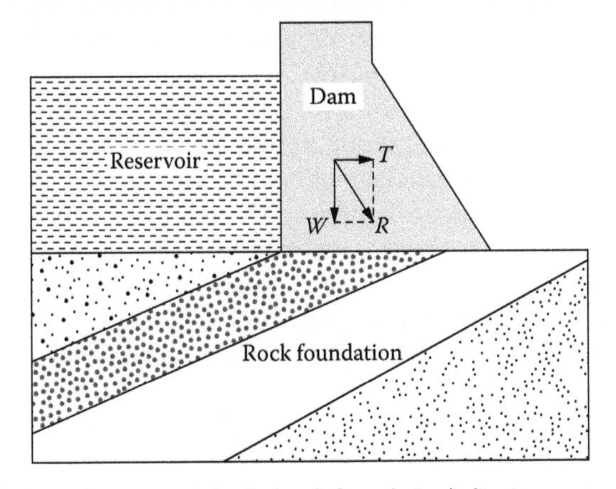

Figure 1.13 A dam resting on rock beds (rock foundation) dipping upstream.

not generally allow the water in the reservoir to percolate below the dam. In fact, the percolating water flows upstream and returns to the reservoir area; thus, the reservoir does not lose much of water due to seepage through the dam foundation. At the same time, the foundation remains watertight and the dam is not subjected to any appreciable amount of uplift pressure.

1.6 WEATHERING OF ROCKS AND SOIL FORMATION

The exposed rocks at the surface of the Earth are subject to continuous decay, disintegration and decomposition under the influence of certain physical, chemical and biological agencies; this phenomenon is called *weathering of rocks*. Temperature variations through a cycle of freezing and thawing of water in the openings inside the rock mass in the cold humid climates and thermal effects in hot dry (arid) regions are responsible for physical/mechanical weathering of rocks. Rainwater causes chemical weathering of rocks because of the chemical action of dissolved atmospheric gases (carbon dioxide, hydrogen, nitrogen etc.). Organisms (burrowing animals, such as earthworms, ants and rodents) and plants also cause degradation of rocks through their physical actions. Human beings also degrade rocks by various activities.

Weathering causes rocks to become more porous, individual grains to be weakened and bonding between mineral grains to be lost. Therefore, rocks lose strength and become more deformable and their permeabilities may change depending on the nature of rocks, the presence and type of weathering and the stage of weathering. The degree of weathering may be reflected by changes in index properties such as dry density, void ratio, clay content and seismic velocity. The engineering suitability of rocks greatly depends on two principal modes of weathering: physical/mechanical weathering (disintegration) and chemical weathering (decomposition). Disintegration of rocks gives rise to satisfactory engineering materials that can be used as pavement materials and concrete aggregates because physical breakdown of the rocks occurs without drastic changes in the rock's minerals and hence without significant reduction in their durability. Decomposition, on the contrary, involves the chemical alteration of the rocks and results in the transformation of most of the important minerals into some form of clay (Weinert, 1974). The assessment of rock weathering has been a challenging problem for engineering geologists and geotechnical engineers. For convenience, rock weathering has been classified into different types/grades by the researchers; Table 1.4 presents the classification suggested by Little (1969).

The processes of soil formation are complex, and they directly affect the engineering properties of the resulting soil mass. Soils are the result of interactions between five soil-forming factors: parent materials, topography,

Table 1.4 Engineering rock weathering classification

Grade	*Degree of weathering*	*Field recognition*	*Engineering properties*
VI	Soil	Surface layer contains humus and plant roots; no recognisable rock structure; unstable on slopes when vegetable cover destroyed	Unsuitable for important foundations; unsuitable on slopes when cover is destroyed
V	Completely weathered	Rock completely decomposed by weathering in place but texture still recognisable; in rock types of granite origin, feldspars completely decomposed to clay minerals; cores cannot be recovered by ordinary rotary drilling methods; can be excavated by hand	Can be excavated by hand or ripping without use of explosives; unsuitable for foundations of concrete dams or large structures; may be suitable for foundations of earth dams and for fill; unsuitable in high cuttings at steep slope angles; requires erosion protection
IV	Highly weathered	Rock so weakened by weathering that fairly large pieces can be broken and crumbled in the hands; sometimes recovered as core by careful rotary drilling; stained by limonite	Similar to grade V; unlikely to be suitable for foundations of concrete dams; erratic presence of boulders makes it an unreliable foundation stratum for large structures
III	Moderately weathered	Considerably weathered; possessing some strength in large pieces (e.g., NX drill cores); often limonite stained; difficult to excavate without use of explosives	Excavated with difficulty without use of explosives; mostly crushes under bulldozer tracks; suitable for foundations of small concrete structures and rockfill dams; may be suitable for semi-pervious fill; stability in cuttings depends on structural features, especially joint attitudes
II	Slightly weathered	Distinctly weathered with slight limonite staining; some decomposed feldspars in granites, strength approaching that of fresh rock; explosives required for excavation	Requires explosives for excavation; suitable for concrete dam foundations; high permeability through open joints; often more permeable than zones above or below; questionable as concrete aggregate
I	Fresh rock	Fresh rock may have some limonite-stained joints immediately beneath weathered rock	Staining indicates water percolation along joints; individual pieces may be loosened by blasting or stress relief and support may be required in tunnels and shafts

Source: Little, A.L., *Proceeding of the 7th International Conference on Soil Mechanics and Foundation Engineering*, Vol. 1, pp. 1–10, 1969.

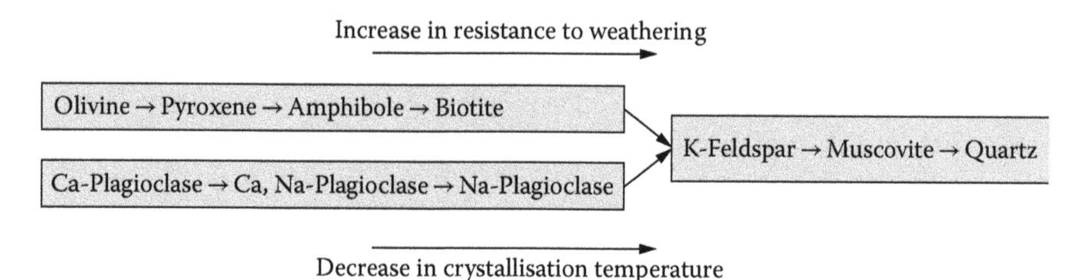

Figure 1.14 Bowen's reaction series.

climate, organisms and time. Weathering of rocks as the parent materials plays a major role in the formation of soils and sediments. Minerals present in the rocks have different degrees of resistance to weathering. Bowen's (1922) reaction series, which lists some minerals in the order of decreasing crystallisation temperature during their formation as a result of cooling of magma, is given in Figure 1.14. This list also follows the order of increasing resistance to weathering after their formation. Olivine, which crystallises earlier, that is, at higher temperature during the formation of rocks from magma, is the most weatherable mineral in rocks. Quartz, which crystallises later, that is, at lower temperature during the formation of rocks from magma, is the least weatherable mineral. Quartz is the most common mineral in soils and sediments as a residue of weathering processes. Weathering of feldspar results in clay minerals (kaolinite or illite). In the tropical weathering environment, the clay minerals break down further, resulting in bauxite and laterite profiles.

It is important to note that a great majority of rocks and soils that are present at or near the ground level are formed during the last one-eighth of their geological time (4600 million years – the age of the Earth and its moon). Approximately seven-eighths of geological history, described as *Precambrian*, is poorly known. Based on the method of formation, soils are basically classified as follows:

1. Sedimentary soils
2. Residual soils
3. Fills
4. Organic soils

Formation of *sedimentary soils* consists of three steps: *sediment formation* as a result of weathering of rocks; *sediment transport* by water, wind, ice, gravity and organisms, called transporting agents and *sediment deposition* in different environments. The orientation and the distribution of particles in a soil mass, commonly termed *soil structure*, is governed by the environment of deposition. There are two types of soil structure, namely the *flocculated structure* and the *dispersed structure* (Figure 1.15). In the

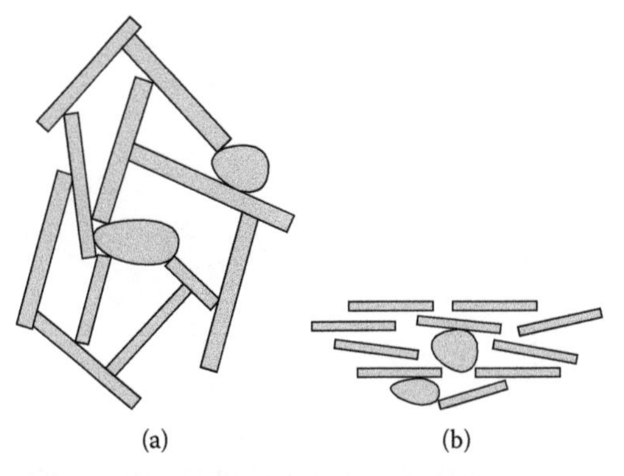

Figure 1.15 Soil structure: (a) flocculated and (b) dispersed.

former structure, the particles have edge-to-face or edge-to-edge contacts, and there is a net attraction, whereas in the latter one, the particles tend to assume a face-to-face orientation, and there is a net repulsion. The engineering behaviour of soil is greatly controlled by the type of structure. In general, an element of flocculated soil has a higher strength, lower compressibility and higher permeability than the element of soil at the same void ratio but in a dispersed state. If the flocculated soil is subjected to a horizontal shear displacement, the particles will tend to line up in the dispersed structure.

Residual soils are products of the in situ weathering of bedrock. The soils are commonly situated above the groundwater table; therefore, they are often unsaturated. *Fill* is a man-made soil; the process of its formation is called *filling*. A fill is actually a sedimentary soil for which man carries out all of the formation processes. *Organic soils* such as peats are derived from the composition of organic materials such as decayed vegetation including leaves and tree roots.

The clay minerals are a group of complex aluminosilicates, mainly formed during the chemical weathering of primary minerals. For example, the clay mineral *kaolinite* is formed by the breakdown of feldspar by the action of water and carbon dioxide. Most clay mineral particles are of 'plate-like' form having a high specific area (surface area of a unit mass of the material), with the result that their properties are influenced significantly more by surface forces than gravitational body forces. Long 'needle-shaped' particles can also occur (e.g., Halloysite) but are rare.

The basic structural units of most clay minerals consist of a *silica tetrahedron* and an *alumina octahedron* (Figure 1.16). The basic units combine to form sheet structures. Silicon and aluminium may be partially replaced by other elements in these units, this being known as *isomorphous substitution*, which may have the following two effects: a net unit charge deficiency results per substitution leading to a net negative charge and a slight distortion of the crystal lattice occurs since the ions are not identical size.

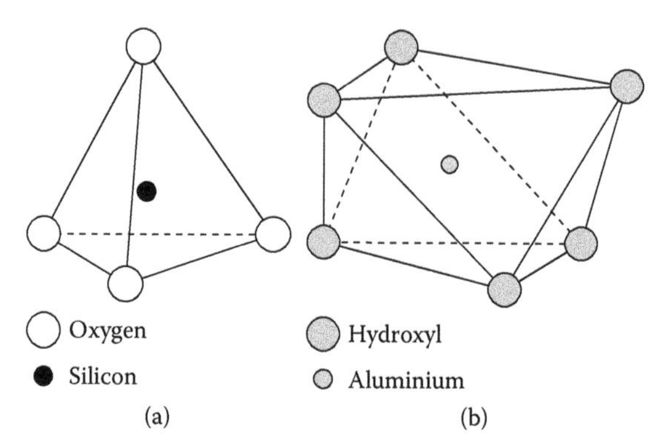

Figure 1.16 Basic structural units: (a) silica tetrahedron and (b) alumina octahedron.

Several clay minerals are formed by the stacking of combinations of the basic sheet structures with different forms of bonding between the combined sheets. There are three principal clay minerals, namely *kaolinite, illite* and *montmorillonite* (Grim, 1968).

The basic structure of kaolinite consists of a layer of alumina octahedron on the top of a layer of silica tetrahedron; this mineral is known as a 'two-layer' mineral. The thickness of the basic unit is about 7.2 Å. There is very limited isomorphous substitution in kaolinite. The combined silica–alumina sheets are held together fairly tightly by hydrogen bonding. A kaolinite particle may consist of over 100 stacks.

The basic structure of illite consists of a sheet of alumina octahedron sandwiched between two sheets of silica tetrahedrons. The thickness of the basic unit is about 10 Å. In the octahedral sheet, there is partial substitution of aluminium by magnesium and iron, and in the tetrahedral sheet, there is partial substitution of silicon by aluminium. The combined sheets are linked together by fairly weak bonding due to non-exchangeable potassium ions held between them.

Montmorillonite has the same basic structure as illite. In the octahedral sheet, there is partial substitution of aluminium by magnesium. The thickness of the basic unit is about 9 Å. The space between the combined sheets is occupied by water molecules and exchangeable cations other than potassium. There is a very weak bond between the combined sheets due to these ions. Considerable swelling of montmorillonite can occur due to addition of water being adsorbed between the combined sheets.

The surfaces of clay mineral particles carry net negative charges, which may arise from any one or a combination of the following five factors: isomorphous substitution, surface disassociation of hydroxyl ions, absence of cations in the crystal lattice, adsorption of anions and presence of organic matter. Out of these five factors, isomorphous substitution of aluminium or silicon atoms by atoms of lower valency is the most important.

A soil particle in nature attracts ions to neutralise its net charge. Since these attracted ions are usually weakly held on the particle surface and can be readily replaced by other ions, they are termed exchangeable ions, and the phenomenon is referred to as *cation exchange*. Calcium is a very common exchangeable ion in soils. The cations are attracted to a clay mineral particle because of the negative surface charges but at the same time tend to move away from each other because of their thermal energy. The net effect is that cations form a dispersed layer adjacent to the particle. The cation concentration decreases with increasing distance from the surface until the concentration becomes equal to that in the normal water in void space. The negatively charged particle surface and the dispersed layer of cations are commonly described as a *double layer*. More details about the double layer can be found in some geotechnical textbooks (e.g., Das, 2013).

1.7 EARTHQUAKES

Earthquakes are vibrations induced in the Earth's crust that virtually shake up a part of the Earth's surface and all the structures and objects lying in that part of the Earth's surface. Earthquakes may or may not result in the actual displacement of a land mass on the Earth's surface. Strong earthquakes are one of the most devastating natural disasters experienced on the Earth (Figure 1.17).

Earthquakes originate due to various causes, which can be classified as *tectonic causes* and *non-tectonic causes*. Tectonic causes include rupture and displacement in the Earth's crustal layers, and they are connected with movement inside the Earth's structure. The earthquakes caused by tectonic

Figure 1.17 The damage done to a road and a house in Sukagawa city, Fukushima prefecture, in northern Japan, 11 March 2011. (Fukushima Minpo/AFP/Getty Images.)

causes are called *tectonic earthquakes*, which are the most common and most destructive events (Figure 1.17). Non-tectonic causes include *natural causes* (large-scale rockfalls or landslides, dashing of sea waves along the coast, waterfalls, natural subsidence such as roof collapse into cavities etc.), *human activities–based causes* (underground nuclear explosions, use of explosives for mineral exploration or excavation works, mining works, movement of heavy trucks and trains, dam construction, deep pumping etc.) and *volcanic causes*. Many of the human activities–based causes result in less energetic earthquakes, but they are important to the engineers because they can cause damage to nearby standing structures and objects. Violent eruption of volcanic lava often causes localised earthquakes. Earthquakes of volcanic origin are less severe and more limited in extent compared to earthquakes caused by tectonic causes.

Unlike most other disasters, earthquakes are nearly impossible to predict. They take place without warning; therefore, people cannot be prepared to save their lives and properties. The science that analyses the causes of earthquakes and the propagation of waves within the Earth and on its surface is called *seismology*.

The causes of tectonic earthquakes are explained by the concept of *plate tectonics*. The basic hypothesis of plate tectonics is that the lithosphere consists of a number of large, intact, rigid blocks called *plates*, which float like large mats on the asthenosphere due to its viscosity and move as a result of convection currents, the force behind plate tectonics. For the study of the causes of the earthquakes, the Earth's crust is divided into six continental-sized plates (African, American, Antarctic, Indo-Australian, Eurasian and Pacific) and about 14 of subcontinental size (e.g., Caribbean, Cocos, Nazca and Philippine).

The point below the ground surface where the rupture of a fault first occurs is the location of origin of the earthquake, which is called the *focus* (or *hypocentre*) of the earthquake (point F in Figure 1.18a). The point vertically above the focus located on the ground surface is called the *epicentre* (point E in Figure 1.18). The vertical distance from the ground surface to the focus is called the *focal depth* (EF in Figure 1.18a). The horizontal distance between the epicentre and a given site is called the *epicentric distance* (EA in Figure 1.18), and the distance between a given site and the focus is called *hypocentric distance* (FA in Figure 1.18a). The intensity of an earthquake decays with the distance. If a line passing through the values of 'same intensity' in a particular earthquake record is imagined on the ground, it is called an *isoseismal line*; several such lines can be imagined.

Based on the focal depth, earthquakes are classified into the following three types:

1. *Deep-focus earthquakes*: These have focal depths of 300–700 km. They constitute about 3% of all the earthquakes recorded around the world.
2. *Intermediate-focus earthquakes*: These have focal depths of 70–300 km.

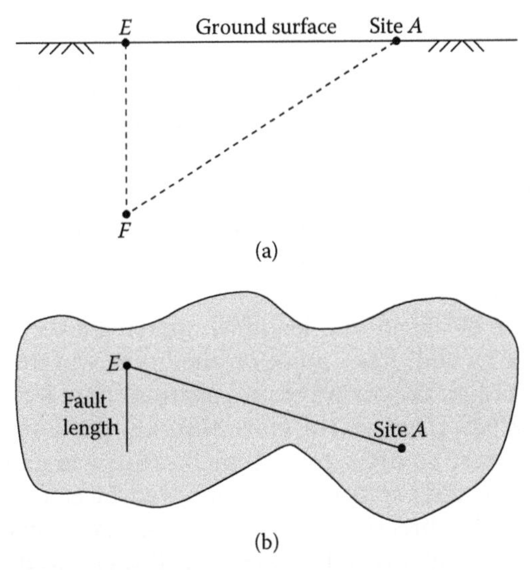

Figure 1.18 Definition of earthquake-related terms: (a) section and (b) plan. (Adapted from Das, B.M. and G.V. Ramana, *Principles of Soil Dynamics*, Cengage Learning, Stamford, CT, 2011.)

3. *Shallow-focus earthquakes*: The focal depths of these earthquakes are less than 70 km. About 75% of all the earthquakes around the world belong to this category.

During the earthquake period, seismic energy generated at the focus propagates in different directions in the form of waves, called *shock* or *seismic waves*. Seismic waves are basically parcels of elastic strain energy that propagate outwards from a seismic source such as an earthquake or an explosion or a mechanical impact. Sources suitable for seismic investigation (see Section 1.9) usually generate short-lived wave trains, known as pulses, which typically contain a wide range of frequencies. Except in the immediate vicinity of the source, the strains associated with the passage of a seismic pulse are small and may be assumed to be elastic. On this assumption, the propagation velocities of seismic pulses are determined by the elastic moduli and densities of the materials through which they pass.

There are two groups of seismic waves: *surface waves* and *body waves*. The surface waves in the form of *Rayleigh waves* and *Love waves* can propagate along the boundary of the solid. Surface waves are felt only near the surface of the Earth when earthquakes occur. They can also travel along the boundary between two media. They play a significant role in the destruction of buildings and other structures during earthquakes. These waves can be observed in a beam by blowing near its side. Rayleigh waves generate a form of swell on the solid surface, whereas Love waves are transverse shear waves on a horizontal surface.

Body waves can propagate through the internal volume of an elastic solid and may be of two types: *compressional waves* (*longitudinal, primary* or *P-waves*), which propagate by compressional and dilational uniaxial strains in the direction of wave travel with particles oscillating about fixed points in the direction of wave propagation and *shear waves* (*transverse, secondary* or *S-waves*), which propagate by a pure shear strain in a direction perpendicular to the direction of wave travel with individual particles oscillating about fixed points in a plane at right angles to the direction of wave propagation.

P-waves can be observed in a beam by applying a compressional stress through striking its end. Each point on the beam vibrates in a sinusoidal movement in the direction of wave propagation, thus a P-wave is a longitudinal wave. When these waves move through the subsurface, they are the first waves perceived after an earthquake. S-waves can be observed in a beam by applying a shear stress to its upper surface. The points oscillate perpendicular to the direction of wave propagation, thus an S-wave is a transverse wave. S-waves travel slower than P-waves and get absorbed in a liquid.

The velocity of a P-wave (V_p) relates to the elastic constants of the medium (bulk modulus of elasticity, K, and shear modulus or modulus of rigidity, G) and its density (ρ) as

$$V_p = \sqrt{\frac{K + \frac{4}{3}G}{\rho}} \tag{1.1}$$

Since K is non-zero for all media (solids, liquids and gases), the P-wave velocity cannot be zero. Therefore, P-waves generated by earthquakes travel in all media and pass through all the layers (crust, mantle and core) of the Earth.

The velocity of an S-wave (V_s) relates only to the shear modulus or modulus of rigidity (G) of the medium and its density (ρ) as

$$V_s = \sqrt{\frac{G}{\rho}} \tag{1.2}$$

Since G is negligible or zero for liquids and gases, the S-wave velocity can be zero. Therefore, S-waves generated by earthquakes travel mainly through solids. Past studies have shown that S-waves do not pass through the outer core of the Earth that extends approximately from 2800 km to 5200 km below the Earth's surface; this observation has indicated that the outer core of the Earth is in a liquid state although the inner core is a solid.

EXAMPLE 1.3

Derive an expression for the ratio of P-wave velocity (V_p) to S-wave velocity (V_s). What do you notice based on this expression?

Solution

From Equations 1.1 and 1.2,

$$\frac{V_p}{V_s} = \sqrt{\frac{3K + 4G}{3G}} = \sqrt{\frac{4}{3} + \frac{K}{G}} \tag{1.3}$$

To calculate V_p/V_s, both K and G of the medium should be known, but this is not essential if the following relationships are used to simplify Equation 1.3:

$$E = 2G(1 + \nu) \tag{1.4}$$

$$E = 3K(1 - 2\nu) \tag{1.5}$$

where E is the Young's modulus of elasticity and ν is the Poisson's ratio.

Eliminating E from Equations 1.4 and 1.5,

$$\frac{K}{G} = \frac{2(1 + \nu)}{3(1 - 2\nu)} \tag{1.6}$$

Using Equation 1.6, Equation 1.3 becomes

$$\frac{V_p}{V_s} = \sqrt{\frac{1 - \nu}{0.5 - \nu}} \tag{1.7}$$

From Equation 1.7, note that the ratio of P-wave velocity to S-wave velocity depends only on the Poisson's ratio of the medium. Thus, by measuring P- and S-wave velocities in the field, one can determine the Poisson's ratio of the rocks and soils at a construction site. Poisson's ratio is an important material parameter for the numerical analysis of Earth slopes and foundations for assessment of their stability.

Since the Poisson's ratio for rocks is typically about 0.25, $\nu_p \approx 1.7\nu_s$, that is, P-waves always travel faster than S-waves in the same medium.

The instrument used to detect and record seismic waves is called a *seismograph*. The basic form of a seismograph contains a heavy weight suspended from a support that is attached to the ground. When waves from an earthquake reach the instrument, the inertia of the weight keeps it stationary while the support attached to the ground vibrates. The movement of the ground in relation to the stationary weight is recorded on a

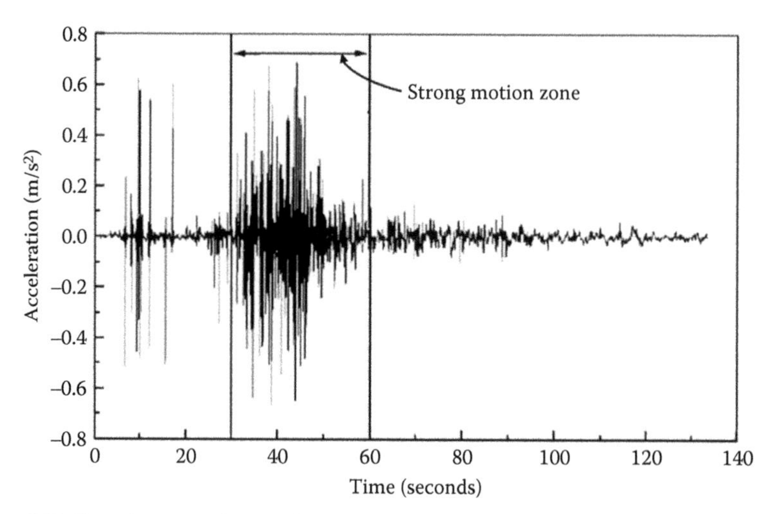

Figure 1.19 Time history of vertical ground acceleration in Bhuj earthquake, 26 January 2001, India. (Adapted from Sitharam, T.G. and L. Govidaraju, *Geotechnical and Geological Engineering*, 22, 439–455, 2004.)

paper wrapped to a rotating drum. Modern seismographs are designed with sensors of the electromagnetic type associated with electronic amplifiers and precision timing system and recorder. Figure 1.19 shows a typical earthquake record, called a *seismogram*. The measurements of seismic waves caused by a particular earthquake at three widely spaced stations can be analysed to work out how far these stations are from the epicentre.

The *intensity of an earthquake* is a measure of the local level of ground shaking as estimated on the basis of human perceptibility and its destructivity. Earthquakes are categorised into 12 grades according to the *Mercalli scale of intensity*, in which grade I refers to the earthquakes that are not felt but can be detected only by instruments, and grade XII refers to the earthquake scenarios that result in situations such as total damage, ground wrapped, waves seen moving through ground and objects thrown upwards. Table 1.5 describes all the 12 grades of the earthquake intensity with peak acceleration values for some grades.

The *magnitude of an earthquake* is a quantitative measure of its size, based on the amplitude of the elastic waves (P-waves) it generates, at known distances from the epicentre. The earthquake magnitude scale presently in use was first developed by C.F. Richter (1958), who summarised its historical developments himself. Richter's earthquake magnitude is defined as

$$\log_{10} E = 11.4 + 1.5M \tag{1.8}$$

Table 1.5 Mercalli maximum earthquake intensity scale

Grades of earthquake intensity	*Damage at epicentrev (and peak acceleration, g being the acceleration due to gravity)*
I	Not felt, detected only by instruments
II	Felt by some persons at rest; suspended objects may swing
III	Felt noticeably indoors; vibration like passing of a truck
IV	Felt indoors by many persons, outdoors by some persons; windows and doors rattle (<0.02g)
V	Felt by nearly everyone; some windows broken
VI	Felt by all, many frightened; some heavy furniture moved; some fallen plaster; general damage slight
VII	Damage to poorly constructed buildings; weak chimneys fall (approx. 0.1g)
VIII	Much damage to buildings, except those specially designed; tall chimneys, columns fall; sand and mud flow from cracks in the ground
IX	Foundations damaged; ground cracked; damage considerable in most buildings; buried pipes broken
X	Disastrous; buildings destroyed; rails bent; small landslides. (>0.6g)
XI	Few structures left standing; wide fissures opened in ground with slumps and landslides
XII	Total destruction; ground wrapped; waves seen moving through ground; objects thrown upwards

where E is the energy released (in ergs) and M is magnitude. Bath (1966) slightly modified the constant given in Equation 1.8 and presented it as

$$\log_{10} E = 12.24 + 1.44M \quad (1.9)$$

From Equation 1.9, it can be seen that an increase of M by one unit generally corresponds to about a 30-fold increase in the energy released (E) due to the earthquake. The smallest felt earthquakes have $M = 2$–2.5, the damaging earthquakes have $M = 5$ or more and any earthquake greater than $M = 7$ is a major disaster.

The length of fault rupture (or fault length) has been found to depend on the magnitude of earthquake. Tocher (1958), based on observations of some earthquakes in the area of California and Nevada, suggested the following relationship:

$$\log_{10} L = 1.02M - 5.77 \quad (1.10)$$

where L is fault length in kilometres.

The Mercalli scale of earthquake intensities and the Richter scale of earthquake magnitudes are not strictly comparable, but $M = 5$ corresponds roughly with grade VI. Table 1.6 presents a rough comparison for other magnitudes and intensities.

Table 1.6 Comparison of Richter magnitude and Mercalli maximum intensity scales

Richter earthquake magnitude scale, M	*Mercalli maximum intensity scale*
1–3	I
3–4	II–III
4–5	IV–V
5–6	VI–VII
6–7	VIII–IX
7–8+	X–XII

1.8 HYDROGEOLOGY

Hydrogeology deals with occurrence, distribution, storage and movement of groundwater in the subsurface. All water below the Earth's surface is referred to as *groundwater* or *subsurface water.* Unlike surface water, groundwater needs very little treatment for use. Groundwater is one of the components of the hydrologic cycle in nature. The groundwater moves slowly through intergranular pores and natural cavities, called *primary openings*, and discontinuities (joints, fractures and solution cavities), called *secondary openings* in rocks. Primary openings are generally found in sedimentary rocks, while secondary openings are found in most igneous and metamorphic rocks and also in some sedimentary rocks. Figure 1.20 gives a physical feel of the presence of pore spaces in a dry weak sandstone sample when immersed into water.

The quantity of groundwater that can be stored in a rock mass depends on its *porosity*, which is defined as a percentage volume of pore spaces/voids/openings in a given rock mass. The property of rock that relates to its ability to transmit water is called *permeability*, which is defined numerically as a flow through unit area of a material in unit time under unit hydraulic head. The SI unit of permeability (also called hydraulic conductivity or coefficient of permeability) is metre per second; for convenience, it is expressed in metre per day for rocks. Table 1.7 gives values of porosity and permeability of some common rocks, soils and rock fracture zones.

All porous rocks are not equally permeable. Permeability of a rock depends on the size of the pore spaces or openings present in the rock and the degree to which they are interconnected. Most soils transmit water through their pores whereas transmission through most rocks is by pores and discontinuities such as joints and fractures. Fractures and joints normally transmit more water than pores. The loads from the structures constructed on the ground can reduce the size of pores and fractures, resulting in reduced permeability. On the contrary, shrinkage due to desiccation can open cracks in clays and dissolution can widen voids in soluble rocks, thus

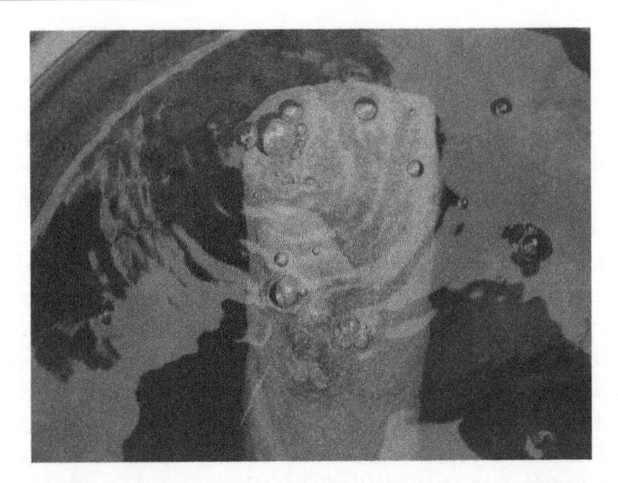

Figure 1.20 Air bubbles from the immersed dry weak sandstone sample collected from the proposed site for the construction of a coal handling plant (CHP), Northern Coalfields Limited, Gorbi, Madhya Pradesh, India. (After Shukla, S.K., Subsoil Investigation for the Estimation of Load-Bearing Capacity of Foundation Soil at the Proposed Site for the Construction of a Coal Handling Plant (CHP), Northern Coalfields Limited, Gorbi, MP, India. A technical report dated 16 October 2006, Department of Civil Engineering, Institute of Technology, Banaras Hindu University, Varanasi, India, 2006.)

Table 1.7 Typical values of porosity and permeability of some common rocks, soils and rock fracture zones

Rocks, soils and rock fracture zones	*Porosity (%)*	*Permeability (m/day)*
Fractured sandstone	15	5
Cavernous limestone	5	Erratic
Shale	3	0.0001
Granite	1	0.0001
Sand	30	20
Gravel	25	300
Clay	50	0.0002
Rock fracture zones	10	50

Source: Waltham, T., *Foundations of Engineering Geology*, Spon Press, London, 2002.

resulting in an increased permeability. Based on the water-bearing and water-yielding properties, geological formations are classified as follows:

Aquifers: Rocks and soils that are both porous and permeable.
Aquicludes: Rocks and soils that are porous but not permeable.
Aquitards: Rocks and soils that are porous but have limited permeability.
Aquifuges: Rocks and soils that are neither porous nor permeable.

Aquifers store groundwater in large quantities and their permeability is to the extent of maintaining a steady supply of a sufficient amount of water to ordinary or pumping wells or springs. The aquifers in which groundwater occurs under atmospheric pressure are called *unconfined aquifers*. If a well is drilled into an unconfined aquifer, the water level in that well represents the *water table*. An aquifer sandwiched between two relatively impermeable strata (aquicludes or aquifuges) is called a *confined aquifer* (also known as an *artesian* or *pressure aquifer*). Since impermeable strata do not allow the movement of groundwater across them, the groundwater within the aquifer remains under pressure greater than atmospheric pressure. The area through which rainwater infiltrates into the confined aquifer is called the *recharge area*. An imaginary surface coinciding with the hydrostatic pressure level of water in the confined aquifer is called the *piezometric surface*. Figure 1.21 shows a typical schematic cross-section of the confined and unconfined aquifers. A well drilled into the ground can be a *water table well* (site A and site B with the well bottom lying in the unconfined aquifer zone), *flowing well* (site B with the well bottom lying in the confined aquifer zone) or *artesian well* (site C with the well bottom lying in the confined aquifer zone), depending on its depth and location on the ground. Thus, in any region, the possibility of obtaining an adequate supply of groundwater is dependent entirely on the location, extent and nature of the aquifers in that region. In river valleys, aquifers occur in abundance and the water table generally lies near the ground. In such regions, the groundwater is usually drawn conveniently through ordinary as well as pumping wells.

Typical aquifers are sand, gravel, sandstone, limestone, grit, conglomerate and so on. The fault-zones, shear-zones, joints and so on, in igneous and metamorphic rocks, may also act as aquifers. For a rock or soil to be an aquifer, its permeability should be greater than 1 m/day (Waltham, 2002).

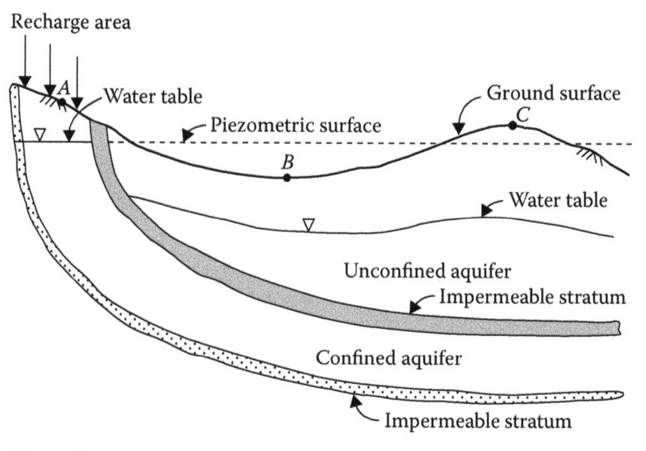

Figure 1.21 Unconfined and confined aquifers.

Clays, shales, mudstones and siltstones are some examples of aquicludes. Clays with restricted amount of silt are aquitards, which are also called *leaky aquifers*. Massive compact granite, syenite, gabbro, gneiss and quartzite without discontinuities are typical examples of aquifuges. These rocks do not allow groundwater to percolate into them at all. For aquicludes and aquifuges, that is, for impermeable rocks or soils, their permeability is generally less than 0.01 m/day (Waltham, 2002).

1.9 SITE INVESTIGATION

Site investigation refers to the appraisal of the surface and subsurface conditions at proposed civil and mining project sites. The engineering geological and geotechnical data and information are required since the planning stage of the project. A typical site investigation includes preliminary studies such as desk study and site reconnaissance, geophysical surveys, drilling boreholes, in situ testing, sampling and laboratory testing of samples and groundwater observations and measurements. *Desk study* involves collection of as much existing information as possible about the site through geological maps, aerial and satellite photographs, soil survey reports, site investigation reports of nearby sites and so on. *Site reconnaissance* consists of a walk-over survey; visually assessing the local conditions such as site access, adjacent properties and structures; topography; drainage and so on.

All the findings of the site investigation are presented to the client in the form of a *site investigation report*, which consists of a site plan, several boring logs that summarise the soil and rock properties at each test pit and borehole and the associated laboratory and in situ test data. The extent of a site investigation program for a given project depends on the type of project, the importance of the project and the nature of the subsurface materials involved. The level of investigation should be appropriate to the proposed site use and to the consequences of failure to meet the performance requirements. For example, a large dam project would usually require a more thorough site investigation than the investigation required for a highway project. A further example is loose sands or soft clays, which usually require more investigation than the investigation required for dense sands or hard clays. The site investigation project can cost about 0.1–1.0% of the total construction cost of the project. The lower percentage is for smaller projects and for projects with less critical subsurface conditions; the higher percentage is for large projects and for projects with critical subsurface conditions.

The purpose of the site investigation is to conduct a scientific examination of the site for collecting as much information as possible at minimal cost about the existing topographical and geological features of the site, for example, the exposed overburden, the course of streams or rivers nearby, rock outcrops, hillocks or valleys, vegetation and mainly the subsurface conditions underlying the site. Investigation of the subsurface conditions at the site for the proposed construction of an engineered system is essential before the design is finalised. Subsurface investigation is needed basically to provide the following:

1. Sequence and extent of each soil and rock stratum underlying the site and likely to be affected by the proposed construction.
2. Engineering geological characteristics of each stratum and geotechnical properties, mainly strength, compressibility and permeability, of soil and rock, which may affect design and construction procedures of the proposed engineered systems and their foundations.
3. Location of groundwater table (or water table) and possible harmful effects of soil, rock and water on materials to be used for construction of structural elements of foundation.

The preceding information is used for determining the type of foundation and its dimensions; estimating the load-carrying capacity of the proposed foundation and identifying and solving the construction, environmental and other potential problems, enabling the civil engineer to arrive at an optimum design with due consideration to the subsurface material characterisation.

Shukla and Sivakugan (2011) have described several methods of subsurface exploration in detail. Experience has shown that making boreholes is the only direct practical method of subsurface exploration to greater depths. Rotary drilling is the most rapid method of advancing boreholes in rock masses unless it is highly fissured; however, it can also be used for all other soils. In this method, cores from rock as well as from concrete and asphalt pavements may be obtained by the use of coring tools (coring bit and core catcher). Coring tools should be designed so that in sound rock, continuous recovery of core is achieved. To obtain cores of the rock, various types of core barrels are available; however, the NX type is commonly used in routine site investigation work, giving core samples of diameter equal to $2\frac{1}{8}$ inch (53.98 mm) (Figure 1.22). It is important to ensure that boulders or layers of cemented soils are not mistaken for bedrock. This necessitates core drilling to a depth of at least 3 m in bedrock in areas where boulders are known to occur. Core photography in colour is performed on all cores to record permanently the unaltered appearance of the rock. Based on the length of rock core recovered from each run, the following quantities may be calculated for a general evaluation of the rock quality encountered:

$$\text{Core recovery} = \left(\frac{\text{Length of the core recovered}}{\text{Total length of the core run}} \times 100\right)\% \tag{1.11}$$

Figure 1.22 Rock cores in a core box.

and

$$\text{Rock quality designation (RQD)} = \left(\frac{\sum \text{lengths of intact pieces of recovered core} \geq 100\,\text{mm}}{\text{Total length of the core run}} \times 100 \right)\% \quad (1.12)$$

A core recovery of 100% indicates the presence of intact rock; for fractured rocks, the core recovery will be smaller than 100%. More details of rock coring and assessment of rock quality are described in Chapter 3.

Geophysical methods can be used to determine the distributions of physical properties, for example, elastic moduli, electrical resistivity, density and magnetic susceptibility at depths below the ground surface that reflect the local subsurface characteristics of the materials (soil, rock or water). These methods may be used for the investigation during the reconnaissance phase of the site investigation programme since it provides a relatively rapid and cost-effective means of deriving aerially distributed information about subsurface stratification. The geophysical investigation can optimise detailed investigation programmes by maximising the rate of ground coverage and minimising the drilling and field testing requirements. Since geophysical investigations may be sometimes prone to major ambiguities or uncertainties of interpretation, these investigations are often verified by drilling or excavating test pits. In fact, geophysical investigation methods may be used to supplement borehole and outcrop data and to interpolate between boreholes.

A wide range of geophysical methods are available for subsurface investigation, for each of which there is an operative physical property to which the method is sensitive (Dobrin, 1976; Kearey et al., 2002). The physical property to which a method responds clearly determines its range of applications. Seismic refraction or reflection and ground-penetrating radar methods can be used to map soil horizons and depth profiles, water tables and depth to bedrock in many situations. Electromagnetic induction, electrical resistivity and induced polarisation (or complex resistivity) methods may be used to map variations in water content, clay horizons, stratification and depth to aquifer/bedrock. The magnetic method is very suitable for locating magnetite and intrusive bodies such as dikes in the subsurface rocks. Other geophysical methods such as gravity and shallow ground temperature methods may be useful under certain specific conditions. Crosshole shear wave velocity measurements can provide soil and rock parameters for dynamic analyses.

Seismic and electrical resistivity methods are routinely used in conjunction with boring logs for subsurface investigation; these methods are therefore described in some detail in Sections 1.9.1 and 1.9.2.

1.9.1 Seismic methods

Seismic methods require generation of shock or seismic waves. They generally use only the P-waves since this simplifies the investigation in two ways. Firstly, seismic/shock detectors that are insensitive to the horizontal motion of S-waves and hence record only the vertical ground motion can be used. Secondly, the higher velocity of P-waves ensures that they always reach a detector before any related S-waves and hence are easier to recognise (Kearey et al., 2002).

Seismic methods make use of the variation of elastic properties of the strata, which affect the velocity of shock/seismic waves travelling through them, thus providing dynamic elastic moduli determinations in addition to the mapping of the subsurface horizons. The required shock waves are generated within the subsurface materials, at the ground surface, or at a certain depth below it, by striking a plate on the soil/rock with a hammer or by detonating a small charge of explosives in the soil/rock. The radiating shock waves are picked up by the vibration detector (e.g., geophone), where the travel times get recorded. Either a number of geophones are arranged in a line or the shock-producing device is moved away from the geophone to produce shock waves at intervals. Figure 1.23 shows the travel paths of primary waves in a simple geological section involving two media (e.g., the soil underlain by bedrock) with respective primary wave velocities of v_1 and v_2 ($>v_1$) separated at a depth z. From the seismic source S, the energy reaches the detector D at the ground surface by three types of ray path.

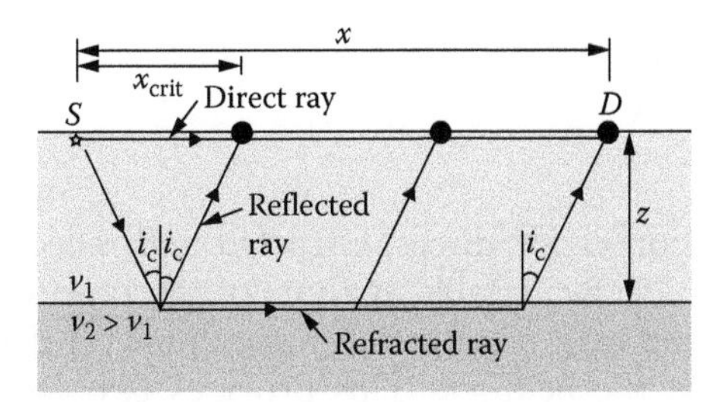

Figure 1.23 Seismic/shock ray paths from a near-surface source to a surface detector in the case of a two-layer system.

The *direct ray* travels along a straight line through the top layer from the source to the detector at velocity v_1. The *reflected ray* is obliquely incident on the interface and is reflected back through the top layer to the detector, having its entire path within the top layer at velocity v_1. The *refracted ray* travels obliquely down to the interface at velocity v_1, along a segment of the interface at the higher velocity v_2, and backs up through the upper layer at velocity v_1.

The travel time t_{dir} of a direct ray is given simply by

$$t_{dir} = \frac{x}{v_1} \tag{1.13}$$

where x is the distance between the source S and the detector D.

The travel time of a reflected ray is given by

$$t_{refl} = \frac{\sqrt{x^2 + 4z^2}}{v_1} \tag{1.14}$$

The travel time of a refracted ray is given by

$$t_{refr} = \frac{z}{v_1 \cos i_c} + \frac{x - 2z \tan i_c}{v_2} + \frac{z}{v_1 \cos i_c} \tag{1.15}$$

where i_c is the critical angle of incidence expressed as

$$i_c = \sin^{-1}\left(\frac{v_1}{v_2}\right) \tag{1.16}$$

Substitution of Equation 1.16 in Equation 1.15 yields

$$t_{refr} = \frac{x}{v_2} + \frac{2z\sqrt{v_2^2 - v_1^2}}{v_1 v_2} \tag{1.17}$$

Time–distance curves for direct, refracted and reflected rays are illustrated in Figure 1.24. By suitable analysis of the time–distance curve for reflected or refracted rays, it is possible to compute the depth to the underlying layer such as the bedrock. This provides two independent seismic methods, namely the *seismic reflection method* and the *seismic refraction method*, for locating the subsurface interfaces. The seismic refraction method is especially useful in determining depth to rock in locations where successively denser strata are encountered, that is, when the velocity of shock or seismic waves successively increases with depth. This method is therefore commonly used in site investigation work. From Figure 1.24, it is evident that the first arrival of seismic energy at a surface detector offset from a surface is always a direct ray or a refracted ray. The direct ray is overtaken by a refracted ray at the *crossover distance* x_{cross}. Beyond this crossover distance, the first arrival is always a refracted ray. Since critically refracted rays travel down to the interface at the critical angle, there is a certain distance, known as the *critical distance* x_{crit}, within which refracted energy will not be returned to the surface. At the critical distance, the travel times of reflected rays and refracted rays coincide because they follow effectively the same path. In the refraction method of site investigation, the detector should be placed at a sufficiently large distance to ensure that the crossover distance is well exceeded so that refracted rays are detected as first arrivals of seismic energy. In general, this approach means that the deeper a refractor, the greater is the range over which recordings of refracted arrivals need to be taken.

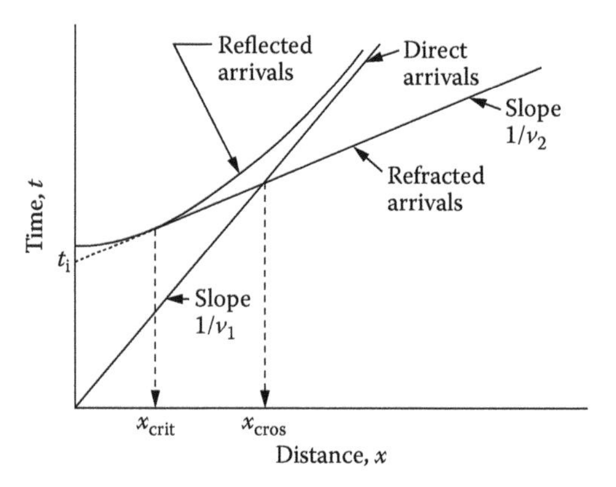

Figure 1.24 Time–distance curves for seismic/shock waves from a single horizontal discontinuity.

In Figure 1.24, the intercept on the time axis of the time–distance plot for a refracted ray, known as the *intercept time* t_i, is given by

$$t_i = 2z\frac{\sqrt{v_2^2 - v_1^2}}{v_1 v_2} \tag{1.18}$$

Since t_i can be determined graphically as shown in Figure 1.24 or numerically from the relation $t_i = t_{refr} - x/v_2$, Equation 1.18 can be used to determine the depth to bedrock as

$$z = \frac{t_i}{2}\frac{v_1 v_2}{\sqrt{v_2^2 - v_1^2}} \tag{1.19}$$

The seismic reflection method may be useful in delineating geological units at depths. Recordings are normally restricted to small offset distances, well within the critical distance for reflecting the interfaces of main interest. This method is not constrained by layers of low seismic velocity and is especially useful in areas of rapid stratigraphic changes.

1.9.2 Electrical resistivity method

The electrical resistivity method may be useful in determining depth to bedrock and anomalies in the stratigraphic profile, in evaluating stratified formations where a denser stratum overlies a less dense medium and in location of prospective sand–gravel or other sources of borrow material. This method is based on the determination of the subsurface distribution of electrical resistivity of earth materials from measurements on the ground surface. Resistivity parameters also are required for the design of grounding systems and cathodic protection for buried structures. The resistivity of a material is defined as the resistance in ohms between the opposite faces of a unit cube of the material. If the resistance of a conducting cylinder having length L and cross-sectional area A is R, the resistivity ρ is expressed in ohm-m (Ω-m) as

$$\rho = R\frac{A}{L} \tag{1.20}$$

The current I is related to the applied voltage V and the resistance R of the material by Ohm's law as

$$I = \frac{V}{R} \tag{1.21}$$

Each soil/rock has its own resistivity depending on water content, compaction and composition. Certain minerals such as native metals and graphite conduct electricity via the passage of electrons. Most of the rock-forming minerals are, however, insulators, and electric current is carried through a rock mainly by the passage of ions in the pore water. Thus, most rocks conduct electricity by electrolyte rather than electronic processes. It follows that porosity is the major control of the resistivity of rocks, and the resistivity generally increases as porosity decreases. However, even crystalline rocks with negligible intergranular porosity are conductive along cracks and fissures. The range of resistivities among earth materials is enormous, extending from 10^{-5} to 10^{15} Ω-m. For example, the resistivity is low for saturated clays and high for loose dry gravel or solid rock as seen in Table 1.8. Since there is considerable overlap in resistivities between different earth materials, identification of a rock is not possible solely on the basis of resistivity data. Strictly, Equation 1.20 refers to electronic conduction but it may still be used to describe the *effective resistivity* of a rock, that is, the resistivity of the soil/rock and its pore water. Archie (1942) proposed an empirical formula for effective resistivity as

$$\rho = a\eta^{-b}S^{-c}\rho_w \tag{1.22}$$

where η is the porosity, S is the degree of saturation, ρ_w is the resistivity of water in the pores and a, b and c are empirical constants. ρ_w can vary considerably according to the quantities and conductivities of dissolved materials.

Normally, one would expect a fairly uniform increase of resistivity with geologic age because of the greater compaction associated with increasing thickness of overburden. There is no consistent difference between the range of resistivities of igneous and sedimentary rocks, although metamorphic rocks appear to have a higher resistivity, statistically, than either of the other rocks (Dobrin, 1976).

Table 1.8 Resistivity of subsurface earth materials

Subsurface earth materials	*Mean resistivity (ohm-m)*
Marble	10^{12}
Quartz	10^{10}
Rock salt	10^{6}–10^{7}
Granite	5000–10^{6}
Sandstone	35–4000
Moraines	8–4000
Limestone	120–400
Clays	1–120

Source: Shukla, S.K. and N. Sivakugan, *Geotechnical Engineering Handbook*, J. Ross Publishing, Inc., Fort Lauderdale, FL, 2011.

The test involves sending direct currents or low-frequency alternating currents into the ground and measuring the resulting potential differences at the surface. For this purpose, four metal spikes are driven into the ground at the surface along a straight line, generally at equal distances; one pair serves as current electrodes, and the other pair as potential electrodes (Figure 1.25). The resistivity can be estimated using the following equation (Kearey et al., 2002):

$$\rho = \frac{2\pi V}{I\left[\left(\frac{1}{r_1}-\frac{1}{r_2}\right)-\left(\frac{1}{R_1}-\frac{1}{R_2}\right)\right]} \tag{1.23}$$

where V is the potential difference between electrodes P_1 and P_2; r_1 and r_2 are the distances from potential electrode P_1 to current electrodes C_1 and C_2, respectively and R_1 and R_2 are the distances from potential electrode P_2 to current electrodes C_1 and C_2, respectively.

When the ground is uniform, the resistivity calculated from Equation 1.23 should be constant and independent of both electrode spacing and surface location. When subsurface inhomogeneities exist, however, the resistivity will vary with the relative positions of the electrodes. Any computed value is then known as the apparent resistivity ρ_a and will be a function of the form of the inhomogeneity. Equation 1.23 is thus the basic equation for calculating the apparent resistivity for any electrode configuration. The current electrode separation must be chosen so that the ground is energised to the required depth and should be at least equal to this depth. This places practical limits on the depths of penetration attainable by normal resistivity methods due to the difficulty in laying long lengths of cable and the generation of sufficient power. Depths of penetration of about 1 km are the limit for normal equipment.

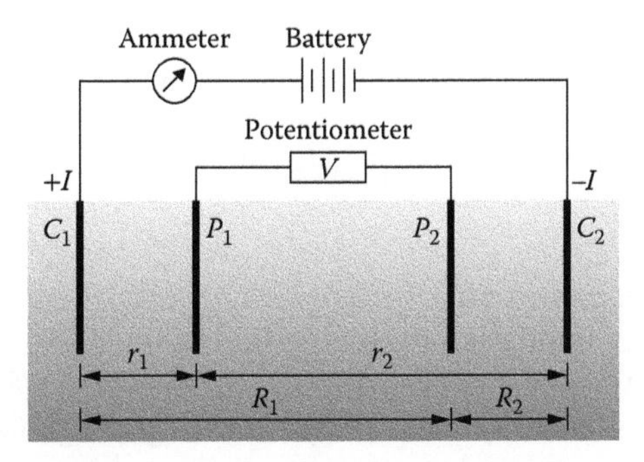

Figure 1.25 The generalised form of the electrode configuration used in the electrical resistivity method (*Note*: C_1 and C_2 are current electrodes, and P_1 and P_2 are potential electrodes).

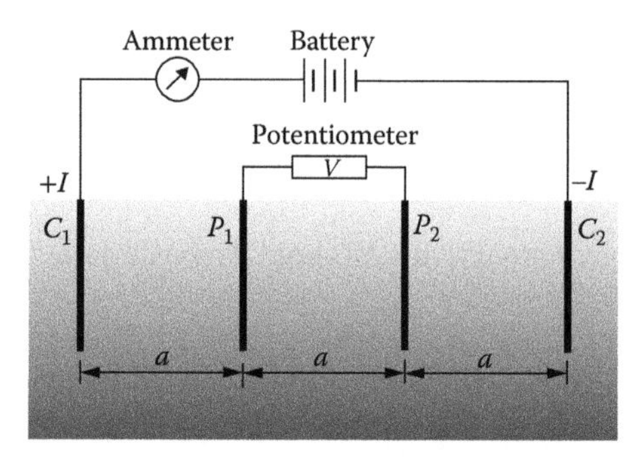

Figure 1.26 The Wenner electrode configuration used in the electrical resistivity method (*Note*: C_1 and C_2 are current electrodes, and P_1 and P_2 are potential electrodes).

There can be several configurations of electrodes, but the *Wenner configuration* is the simplest in that current and potential electrodes are maintained at an equal spacing a (see Figure 1.26). Substitution of this condition, that is, $r_1 = a$, $r_2 = 2a$, $R_1 = 2a$ and $R_2 = a$ in Equation 1.23 yields

$$\rho_a = 2\pi a \frac{V}{I} \tag{1.24}$$

In the study of horizontal or near-horizontal overburden soil–bedrock interfaces, the spacing a is gradually increased about a fixed central point. Consequently, readings are taken as the current reaches progressively greater depths. This technique, known as *vertical electrical sounding* (VES), also called *electrical drilling* or *expanding probe*, is extensively used to determine overburden thickness and also to define horizontal zones of porous media. To study the lateral variation of resistivity, the current and potential electrodes are maintained at a fixed separation and progressively moved along a profile. This technique, known as constant separation traversing (CST) (also called *electrical profiling*), is used to determine variations in bedrock depth and the presence of steep discontinuities.

1.10 SUMMARY

1. *Soils* and *rocks* are made up of small crystalline units known as minerals, and they constitute the crust (thin outer shell) of the Earth. The crust has a thickness of 30–35 km in continents and 5–6 km in oceans.
2. A mineral is basically a naturally occurring inorganic substance composed of one or more elements with a unique chemical composition and arrangement of elements (crystalline structure) and distinctive physical properties (colour, streak, hardness, cleavage, fracture, lustre, habit,

tenacity, specific gravity, magnetism, odour, taste and feel). Hardness of minerals increases from 1 (for talc) to 10 (for diamond). Silica (quartz) and feldspars are the most common rock-forming minerals.

3. A rock is a hard, compact, naturally occurring earth material composed of one or more minerals and is permanent and durable for engineering applications. Rocks generally require blasting for their excavation, and they are classified as igneous rocks (granite, basalt, rhyolite etc.), sedimentary rocks (sandstone, limestone, conglomerate etc.) and metamorphic rocks (quartzite, slate, marble etc.). Compared to igneous and metamorphic rocks, sedimentary rocks are less strong and durable. In nature, one type of rock changes slowly to another type, forming a rock cycle. All kinds of rocks in the form of dressed blocks or slabs, called building stones, or in any other form, called building materials, are frequently used in civil engineering projects.
4. The strike of a plane is the direction of a line considered to be drawn along the plane so that it is horizontal. The line of maximum inclination on the inclined plane is called the line of dip. Dip direction (or dip azimuth) is the direction of the horizontal trace of the line of dip, expressed as an angle measured clockwise from the north.
5. Folds are defined as wavy undulations developed in the rocks of the Earth's crust. Faults are fractures in crustal strata along which appreciable shear displacement of the adjacent rock blocks have occurred relative to each other, probably due to tectonic activities. Discontinuity is a collective term used for all structural breaks (bedding planes, fractures and joints) in solid geologic materials that usually have zero to low tensile strength.
6. The exposed rocks at the surface of the Earth are subject to continuous decay, disintegration and decomposition under the influence of certain physical, chemical and biological agencies; this phenomenon is called weathering of rocks. Weathering causes rocks to become more porous, individual grains to be weakened and bonding between mineral grains to be lost.
7. Soils are the result of interactions between five soil-forming factors: parent materials, topography, climate, organisms and time. Quartz that crystallises later, that is, at lower temperature during the formation of rocks from magma, is the least weatherable mineral present in soils and rocks.
8. Earthquakes are vibrations induced in the Earth's crust that virtually shake up a part of the Earth's surface and all the structures and objects lying in that part of the Earth's surface. About 75% of all the earthquakes around the world have been shallow-focus earthquakes (focal depths less than 70 km). There are two groups of seismic waves: surface waves (Rayleigh waves and Love waves) and body waves (P-waves and S-waves). P-waves always travel faster than S-waves in the same medium.

9. The instrument used to detect and record seismic waves is called a seismograph, and the records are called seismograms. The intensity of an earthquake is a measure of the local level of ground shaking as estimated on the basis of human perceptibility and its destructivity. The magnitude of an earthquake is a quantitative measure of its size, based on the amplitude of elastic waves (P-waves) it generates, at known distances from the epicentre. The smallest felt earthquakes have magnitude $M = 2–2.5$, the damaging earthquakes have $M = 5$ or more and any earthquake greater than $M = 7$ is a major disaster.
10. Aquifers store groundwater in large quantities and their permeability is to the extent of maintaining a steady supply of a sufficient amount of water to ordinary or pumping wells or springs. Typical aquifers are sand, gravel, sandstone, limestone, grit, conglomerate and so on. The fault-zones, shear-zones, joints and so on, in igneous and metamorphic rocks, may also act as aquifers.
11. A typical site investigation includes preliminary studies such as desk study and site reconnaissance, geophysical surveys, drilling boreholes, in situ testing, sampling and laboratory testing of samples and groundwater observations and measurements. All the findings of the site investigation are presented to the client in the form of a site investigation report, which consists of a site plan, several boring logs that summarise the soil and rock properties at each test pit and borehole and the associated laboratory and in situ test data.
12. Rotary drilling is the most rapid method of advancing boreholes in rock masses unless it is highly fissured. In bedrocks, core drilling to a depth of at least 3 m should be done. The geophysical investigation can optimise detailed investigation programmes by maximising the rate of ground coverage and minimising the drilling and field testing requirements.

Review Exercises

Select the most appropriate answers to the following 10 multiple-choice questions:

1. The difference in equatorial and polar radii of the Earth is approximately
   a. 0 km
   b. 22 km
   c. 44 km
   d. 122 km
2. A acidic igneous rock has
   a. A definite chemical composition
   b. No definite chemical composition
   c. Silica content greater than 60%
   d. Both (b) and (c)

3. On Mohs scale, the hardness of orthoclase is
   a. 2
   b. 4
   c. 6
   d. 8
4. The ratio of the apparent dip to the true dip for a given bedding plane is
   a. Equal to 1
   b. Greater than 1
   c. Equal to or greater than 1
   d. Less than 1
5. Select the incorrect statement.
   a. In anticlines, the older rock beds generally occupy a position in the interior (core) of the curvature.
   b. The discontinuities in rocks make them anisotropic.
   c. On a geological map, dashed lines represent the boundaries between the outcrops of rock beds.
   d. None of the above.
6. Which of the following minerals is the most weatherable?
   a. Quartz
   b. Olivine
   c. Feldspar
   d. Pyroxene
7. Which grade of the Mercalli scale of earthquake intensities corresponds roughly with the Richter earthquake magnitude $M = 5$?
   a. II
   b. IV
   c. VI
   d. VIII
8. S-waves pass through
   a. Solids
   b. Only solids
   c. Liquids
   d. Both solids and liquids
9. Rocks and soils having porosity but limited permeability are called
   a. Aquifers
   b. Aquicludes
   c. Aquitards
   d. Aquifuges
10. For site investigation purposes, the minimum depth of core drilling in bedrock is
    a. 1 m
    b. 3 m
    c. 6 m
    d. 9 m
11. What is the difference between the continental crust and the oceanic crust?

12. How does the temperature vary within the Earth?
13. What are minerals? Enumerate the physical properties of minerals. Are coal and petroleum minerals?
14. How are minerals identified? Explain two common methods. How do you determine the hardness of a mineral?
15. What are the rock-forming processes? Explain the different types of rocks with some typical examples? How are rocks distinguished from each other?
16. What do you mean by rock cycle? Explain with the help of a neat sketch.
17. List the essential rock-forming minerals. Indicate the minerals common in igneous rocks.
18. Classify the following rock types in terms of igneous, sedimentary and metamorphic and indicate important minerals in each of them: granite, quartzite, basalt, sandstone, marble and limestone.
19. Name three metamorphic rocks and indicate the original rock prior to metamorphism in each case.
20. Define strike and dip. Also define other terms used to describe the orientation of a rock bed, discontinuity plane or sloping ground and explain them with the help of a neat sketch.
21. What are folds? Explain the different elements of a fold with the help of a neat sketch.
22. What are the differences between an anticline and a syncline? Explain briefly with the help of a neat sketch.
23. What are faults? Describe the different elements of the fault with the help of a neat sketch.
24. What are joints and how do they differ from faults and fractures?
25. What is an unconformity? What does it signify? What is the most common rock type that is present at an unconformity? How is it presented in a geological map?
26. How are geological structures in rocks important in civil engineering practice? Explain briefly.
27. What is weathering? Describe the different processes of weathering.
28. Arrange the rock-forming minerals in an increasing sequence of their resistance to weathering.
29. What are the different grades of rock weathering?
30. How do soils form? What are clay minerals? Explain the different types of soil structure that affect their engineering behaviour.
31. What are the civil engineering considerations of weathering products?
32. What are earthquakes? Give an account of their main causes and effects. How are earthquakes classified?
33. Enumerate the different earthquake-related terms and explain them.
34. What are the differences between P-waves and S-waves? How does their ratio vary with the Poisson's ratio of soils and rocks?

35. What is the difference between the intensity and magnitude of an earthquake? How are they defined and classified?
36. Discuss the effects of discontinuities on the selection of sites for dam projects.
37. What are the special qualities of rocks that make them suitable for building stones and materials?
38. Define porosity. How is permeability different from porosity?
39. What are aquifers? What are their different types? Explain with the help of a neat sketch.
40. How is an aquifer different from an aquiclude, aquitard and aquifuge? Give two examples for each of them.
41. Explain the basic objectives of site investigation? List the methods of site investigation.
42. What are the methods of surface and subsurface exploration for foundation rocks? Explain them briefly.
43. Explain the factors affecting the electrical resistivity of earth materials?
44. Explain the procedure for the estimation of thickness of earth formations using the seismic refraction technique.

Answers:
1. b; 2. d; 3. c; 4. d; 5. c; 6. b; 7. c; 8. b; 9. c; 10. b

REFERENCES

Anson, R.W.W. and Hawkins, A.B. (1999). Analysis of a sample containing a shear surface from a recent landslip, south Cotswolds, UK. *Geotechnique*, Vol. 49, No. 1, pp. 33–41.

Archie, G.E. (1942). The electrical resistivity log as an aid in determining some reservoir characteristics. *Transactions of the American Institute of Mining and Metallurgical Engineers*, Vol. 146, pp. 54–52.

Bath, M. (1966). Earthquake seismology. *Earth Science Reviews*, Vol. 1, p. 69.

Bowen, N.L. (1922). The reaction principles in petrogenesis. *Journal of Geology*, Vol. 30, pp. 177–198.

Das, B.M. (2013). *Fundamentals of Geotechnical Engineering*. 4th edition, Cengage Learning, Stamford.

Das, B.M. and Ramana, G.V. (2011). *Principles of Soil Dynamics*. Cengage Learning, Stamford.

Dobrin, M.B. (1976). *Introduction to Geophysical Prospecting*. McGraw-Hill Book Company, Inc., New York.

Grim, R.E. (1968). *Clay Mineralogy*. 2nd edition, McGraw-Hill Book Co, Inc., New York.

Kearey, P., Brooks, M. and Hill, I. (2002). *An Introduction to Geophysical Exploration*. Blackwell Science Ltd., London.

Little, A.L. (1969). The engineering classification of residual tropical soils. *Proceeding of the 7th International Conference on Soil Mechanics and Foundation Engineering*, Mexico, Vol. 1, pp. 1–10.

Mukerjee, P.K. (1984). *A Textbook of Geology*. The World Press Private Limited, Kolkata.

Raymahashay, B.C. (1996). *Geochemistry for Hydrologists*. Allied Publishers Ltd., New Delhi.

Richter, C.F. (1958). *Elementary Seismology*. W.H. Freeman, San Francisco, CA.

Shukla, S.K. (2006). Subsoil Investigation for the Estimation of Load-Bearing Capacity of Foundation Soil at the Proposed Site for the Construction of a Coal Handling Plant (CHP), Northern Coalfields Limited, Gorbi, MP, India. A technical report dated 16 October 2006, Department of Civil Engineering, Institute of Technology, Banaras Hindu University, Varanasi, India.

Shukla, S.K. (2007). Allowable Bearing Pressure for the Foundation Rock/Soil at km 0.080 of the Meja-Jirgo Link Canal for the Proposed Construction of a Hydraulic Structure (Head Regulator), Mirzapur, UP, India. A technical reported dated 11 June 2007, Department of Civil Engineering, Institute of Technology, Banaras Hindu University, Varanasi, India.

Shukla, S.K. and Sivakugan, N. (2011). Site Investigation and In Situ Tests, in *Geotechnical Engineering Handbook*, Das, B.M., editor, J. Ross Publishing, Inc., Fort Lauderdale, FL, pp. 10.1–10.78.

Sitharam, T.G. and Govidaraju, L. (2004). Geotechnical aspects and ground studies in Bhuj earthquake, India. *Geotechnical and Geological Engineering*, Vol. 22, pp. 439–455.

Tocher, D. (1958). Earthquake energy and ground breakage. *Bulletin, Seismological Society of America*, Vol. 48, No. 2, pp. 147–153.

Waltham, T. (2002). *Foundations of Engineering Geology*. 2nd edition, Spon Press, London.

Weinert, H.H. (1974). A climatic index of weathering and its application in road construction. *Geotechnique*, Vol. 24, No. 4, 475–488.

Chapter 2

Spherical presentation of geological data

2.1 INTRODUCTION

Construction activities on rocks have been reported several centuries ago. The applications include foundations, slope stability, underground excavations and so on. The early activities include structures and monuments built in Greece, Egypt, Iraq, India and China. Some of the notable examples are the Pyramids of Giza (twenty-sixth century BC), Abu Simbel temple (twelfth century BC), Hanging Gardens of Babylon (sixth century BC) and Parthenon (fifth century BC). Figure 2.1a shows the Parthenon temple on the Acropolis Hill, Athens, Greece. It was built of marble, on a limestone hill that was underlain by phyllites. Figure 2.1b shows the 6300-m-long Corinth Canal in Greece, which has a depth varying up to 78 m. The Corinth Canal project started a few centuries BC and was abandoned and later completed only in 1893.

The term *rock mass* applies to a large extent of rock, from several metres to few kilometres, which can include many *discontinuities* of different forms. The presence of discontinuities such as faults, joints and bedding planes in the rock mass as described in Section 1.5 influences its engineering behaviour.

Our ability to present the orientations of the various discontinuities and their intersections and interpret them is a prerequisite for carrying out a proper analysis of the rock mass behaviour. With the fundamentals of engineering geology covered in Chapter 1, we will continue the introduction to rock mechanics through this chapter on spherical presentation of geological data, which is a systematic method of presentation (e.g., orientation of the discontinuity planes as introduced in Section 1.5) that enables some simple stability analyses in engineering applications to be carried out.

2.2 ORIENTATIONS OF PLANES AND LINES

In rock mechanics and geology, we deal with *discontinuities*, which include bedding planes, faults and joints. It is very important to define the orientation of these planes without any ambiguity. Some of the common terms

(a)

(b)

Figure 2.1 (a) Parthenon temple on Acropolis Hill, Athens, Greece and (b) Corinth Canal, Greece.

associated with the orientation of a plane are *dip* (ψ), *dip direction* (α) and *strike*. Dip, also known as the *true dip*, is the steepest inclination of the plane to horizontal. *Apparent dip* is the inclination of any arbitrary line on the plane to horizontal, which is always less than the true dip. When a marble is rolled down the plane, it follows the line of maximum inclination, defining the true dip. Strike is the *trace* (or intersection) of the dipping plane with the horizontal reference plane. It is also the orientation of the horizontal line drawn on the dipping plane. It is perpendicular to the dip direction.

Figure 2.2 shows an inclined plane (dark grey) for which we will define the dip and the dip direction. The inclined plane intersects the horizontal plane (light grey) along a horizontal line, which is known as the strike. The direction

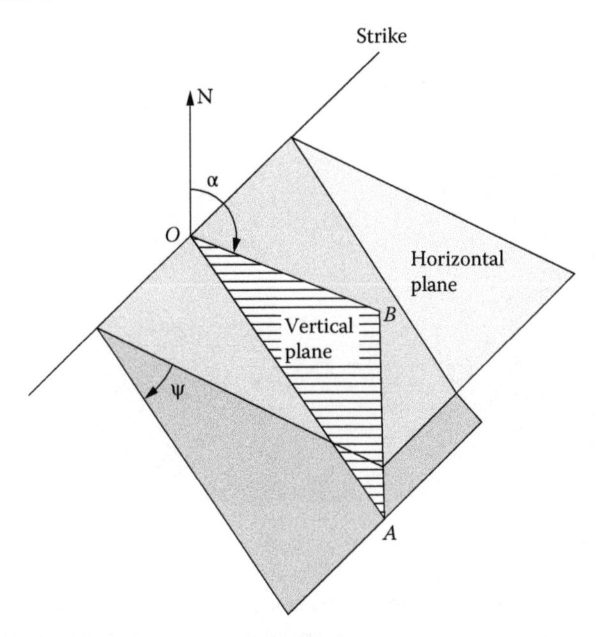

Figure 2.2 Definition of dip and dip direction.

of the strike can be specified as N30E, for example, implying the line is at 30° to north on the eastern side. This is the same as S30W. If a marble is dropped from point O, it will travel along the steepest line *OA* on the slope, known as the *line of dip*, which is always perpendicular to the strike. Let us consider the vertical plane through *OA*, which intersects the horizontal plane along the line *OB*. The direction of *OB* with respect to north is the dip direction (α), which can be in the range of 0–360°. A plane dipping towards east has a dip direction of 90°. Dip direction, also known as *dip azimuth*, is the direction of the horizontal trace (projection) of the line of dip, measured clockwise from the north. Dip (ψ) is the angle the inclined plane makes with the horizontal, which is in the range of 0–90°. A horizontal plane has dip of 0° and a vertical plane has dip of 90°. A plane can be specified as 40/210, 20/080 and so on, where the angles before and after the slash denote the dip and the dip direction, respectively. The dip is specified in two digits and the dip direction in three digits to avoid confusion. Sometimes in literature, these two angles are interchanged, thus giving the dip direction first, followed by the dip. In the field, dip and dip direction of a plane can be measured by a geological compass shown in Figure 2.3. The measurement technique is fairly straightforward. The large horizontal dial is a compass that reads the dip direction, and the small vertical dial reads the dip.

When dealing with the axis of a borehole or a tunnel, or the intersection of two planes, we are dealing with *lines* and not planes. The orientation of a line is defined by *plunge* and *trend*. The plunge of a line (similar to dip of a

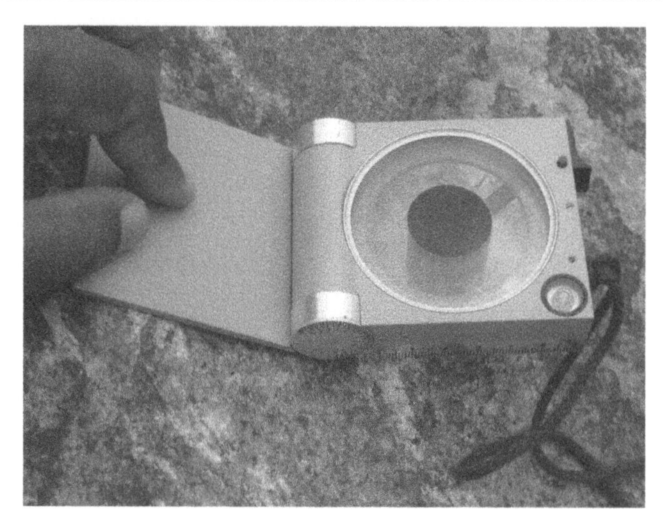

Figure 2.3 Geological compass.

plane) is the inclination of the line to horizontal. It is taken as positive when the line is below horizontal and negative when above horizontal. The trend (similar to dip direction) is the direction of the horizontal projection (or trace) of the line, measured clockwise from the north. The symbols (ψ and α) and the ranges are the same as before.

2.3 COORDINATE SYSTEM WITH LONGITUDES AND LATITUDES

Spherical projections are used to graphically represent geological data such as the orientations of the bedding planes and other discontinuities. We look at Earth as having *longitudes* (or *meridians*) and *latitudes* as shown in Figure 2.4, which are used to locate a point on the globe. It is in fact a 'coordinate system' that we will use in rock mechanics too, but with some modification. Meridians and latitude lines are perpendicular to each other.

Let us consider a *reference sphere* shown in Figure 2.4, which will be used as the basis for the stereographic projection study. A peripheral circle for which the centre coincides with that of the reference sphere is known as a *great circle*. It is formed at the intersection of a *diametric plane* and the sphere. It can be in any orientation and there are thousands of great circles. Each line of longitude (or meridian) passes through the North and the South Poles and hence is part of a great circle. The *equator* is a line of latitude that divides the sphere along the *equatorial plane* into two halves – upper and lower hemispheres. The equator is a great circle that corresponds to 0° latitude. All other lines of latitudes are not great circles; they are *small circles* (see Figure 2.5). They are literally smaller than the great circles.

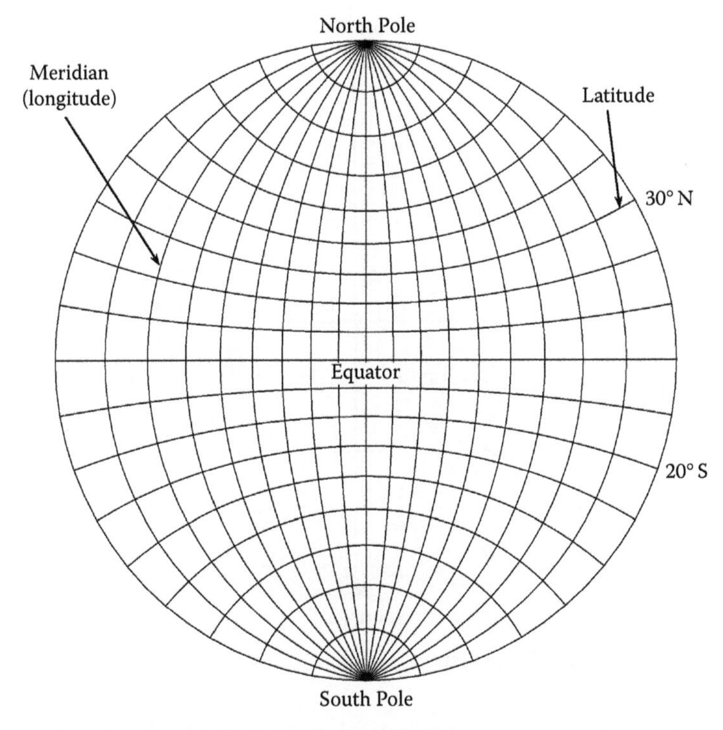

Figure 2.4 Reference sphere, longitudes and latitudes.

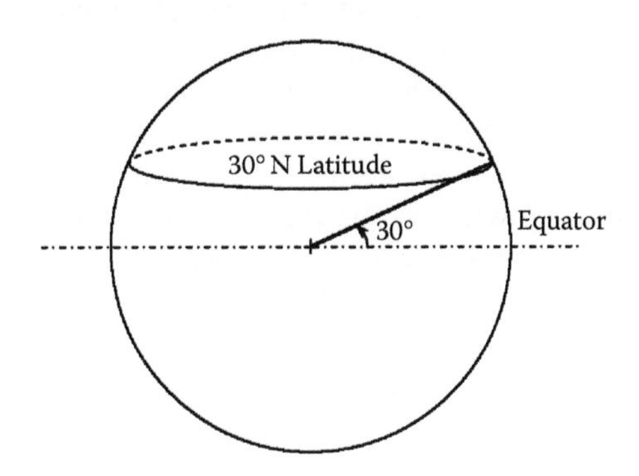

Figure 2.5 Definition of line of latitude.

What does a 30°N line of latitude mean? A radial line to any point on the 30° line of latitude subtends 30° to horizontal at the centre, as shown in Figure 2.5. The equator is taken as the reference line for assigning latitudes as 30°N, 20°S and so on. In the northern and southern hemispheres, latitudes can be in the range of 0–90°. In a similar manner, we have to select one of the longitudinal lines as the reference line and give longitude of a point on the sphere as an angle in the range of 0–360° from this line. Remember, the longitudinal line

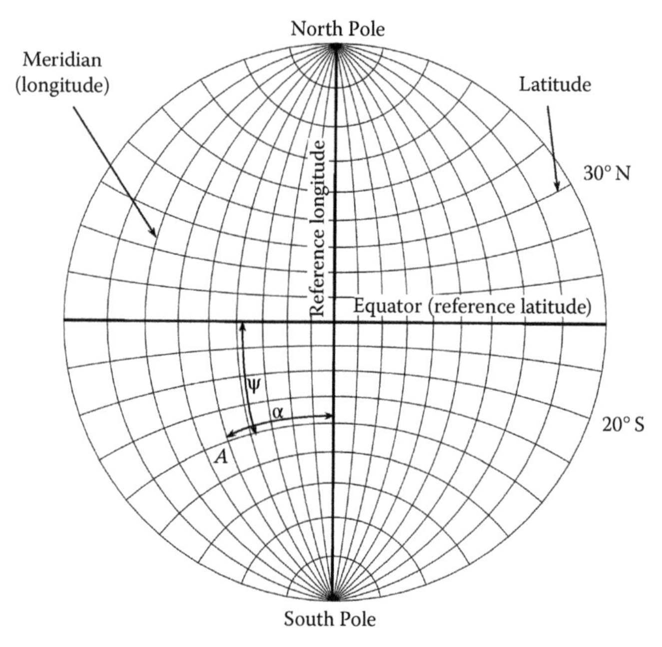

Figure 2.6 A simple coordinate system.

that passes through Greenwich, England, is the reference line for defining the longitude of a city. In the Global Positioning System (GPS) in your car, you see these two values that define your current location on the globe.

In spherical projections, we have a simple coordinate system for locating a point (A) on a sphere. The coordinates are the latitude and the longitude, with the ranges of 0–90° (considering only one half) and 0–360°, respectively. We have also shown that the dip (ψ) and the dip direction (α) of a plane have the same ranges. Let us see how we can use this coordinate system to represent the dip and the dip direction of a plane by the latitude and the longitude, respectively (Figure 2.6). We will be mainly using the lower half of the sphere below the equator.

2.4 INTERSECTION OF A PLANE AND A SPHERE

Figure 2.7 shows a plane passing through the centre of a *reference sphere* where the intersection is shown as a dark shaded circle. Such a circle, where the centre coincides with that of the sphere, is a great circle. The second horizontal great circle shown as a dashed line is the one that separates the reference sphere into upper and lower halves. The lower half and the upper half of the reference sphere represent the same information about the plane, and hence from now on, we will only refer to the *lower reference*

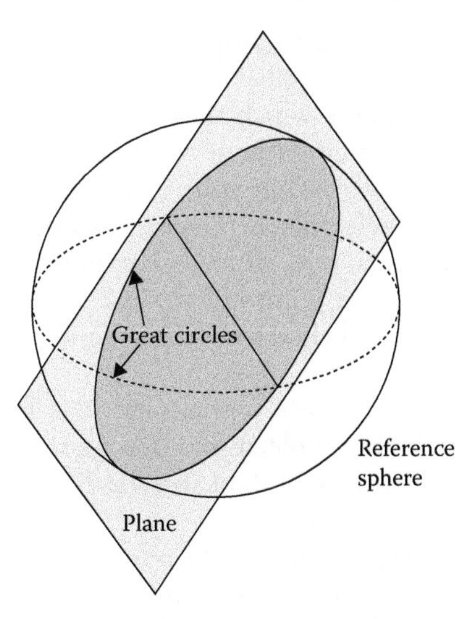

Figure 2.7 Intersection of a plane with reference hemisphere.

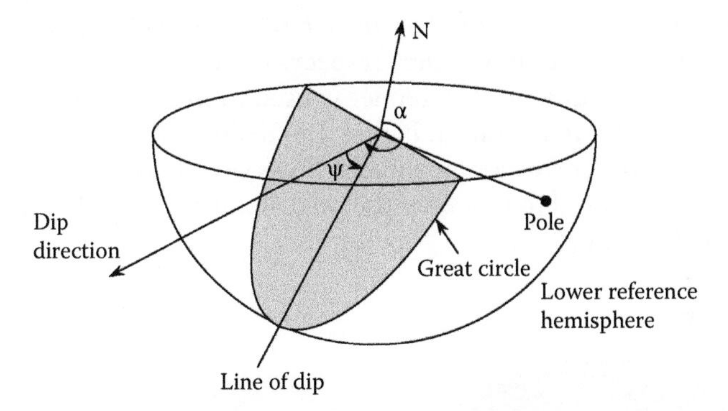

Figure 2.8 Intersection of the dipping plane with the lower reference hemisphere.

hemisphere for simplicity. The intersection of the plane with the lower half of the reference hemisphere is shown in Figure 2.8. It can be seen that any plane with a specific dip (ψ) and dip direction (α) can be graphically presented using a reference lower hemisphere shown in Figure 2.8. For every plane, the intersection creates a unique great circle, representing the specific values of ψ and α. Let us see how we can do this a bit more systematically and present this quantitatively.

Let us imagine that the lower reference sphere is shifted, without any rotation, to the location of the dipping plane of interest. The shifting is purely translational – that is, north remains north. The lower hemisphere is shifted until the plane passes through the centre of the hemisphere, making the plane diametrical (see Figures 2.7 and 2.8). The intersection of the plane and the *lower reference hemisphere* will define a unique great circle that reflects the dip and the dip direction of the plane that is in a three-dimensional space. This applies to lines too, which are represented by points on the reference sphere which they pierce through. Here, the lower reference hemisphere is shifted without any rotation until the line of interest passes through the centre of the sphere. The intersection of the line at the surface of the lower reference hemisphere is known as the *pole of the line*. In Figure 2.8, the radial line, which is also normal to the plane, pierces the reference hemisphere at a point known as the *pole of the plane*. Every plane has a unique pole and therefore, the pole itself can also be used to represent a plane.

Planes in three dimensions are represented in a lower reference hemisphere by a great circle or a pole. Lines are represented only by a pole. To present the three-dimensional data in two dimensions, the concept of spherical projections comes in handy.

The front view and the plan view of the reference sphere in Figure 2.4 are shown in Figure 2.9. The first (Figure 2.9a) is known as the *equatorial stereonet* or *meridional stereonet* showing the two-dimensional projections of the longitudes, and the latitudes. The second (Figure 2.9b) is the *polar stereonet*, showing a series of concentric circles and radial lines, which are the same latitudes and longitudes, respectively when projected on a horizontal plane. The equatorial stereonet is used to present the projections of great circles similar to the one in Figure 2.8, defining the dip (ψ) and the dip direction (α) of the corresponding plane. Poles can also be shown there. The polar stereonet is used to plot only the pole of a plane, which is adequate to fully define the plane.

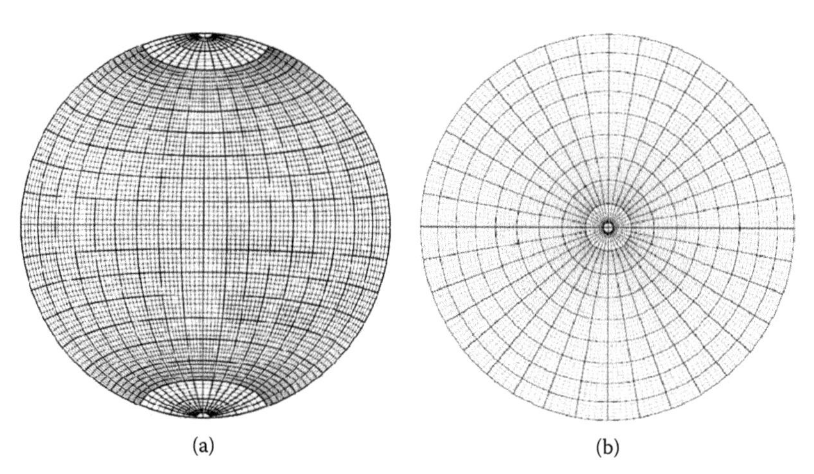

Figure 2.9 Stereonets: (a) front view – equatorial projection and (b) plan view – polar projection.

2.5 SPHERICAL PROJECTIONS

We all know of plan view in engineering drawings, which we have seen in building plans, site layouts and so on. There are a few other ways of projecting the points on the surface of a sphere onto a horizontal or vertical plane. Two of the common methods of projection are (1) *equal area projection* and (2) *equal angle projection*. Both methods are good, but they should never be mixed because they are different. Analysing data originally plotted on an equal area net using an equal angle net or vice versa can lead to erroneous interpretations. Always note the type of projection used and avoid any confusion. *From now on, we will use equal area projection.*

We are gradually developing the skills of imagining the field situation in a three-dimensional space, which we can only present in two dimensions. This is a very important skill for mastering spherical projections.

2.5.1 Equal area projection

Equal area projection is sometimes called *Lambert* or *Schmidt* projection. The point A in Figure 2.10 is projected to A' on the horizontal plane at the bottom of the hemisphere, by swinging an arc centred at O, the point of contact between the sphere and the horizontal plane (i.e., distance $OA = OA'$). This way, every point on the lower hemisphere can be mapped onto a unique point on the horizontal plane. The furthest point from O is on the horizontal rim of the hemisphere, at a distance of $\sqrt{2}R$ and hence the projection area will have a radius of $\sqrt{2}R$. Think of peeling half an orange and leaving the skin on the table flat – it is similar. The surface area of the lower hemisphere is $2\pi R^2$. This is mapped onto an equal area on the horizontal plane in the form of a circle that has radius of $\sqrt{2}R$. An area on the lower hemisphere will have the same area when projected onto the horizontal plane, without any distortion, hence the name equal area projection. The equatorial and polar projections in Figure 2.9 are equal area projections.

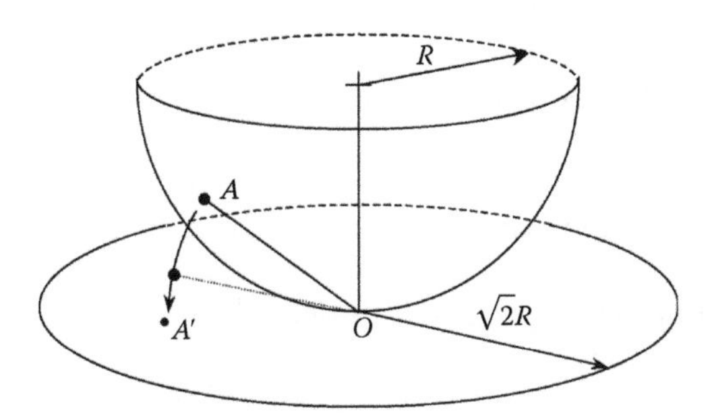

Figure 2.10 Equal area projection method.

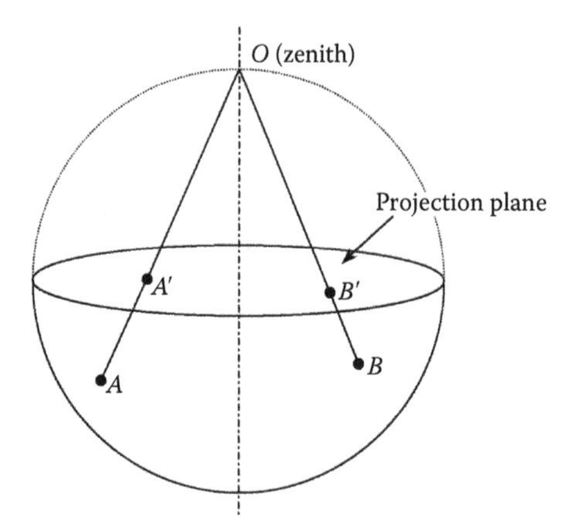

Figure 2.11 Equal angle projection method.

2.5.2 Equal angle projection

Equal angle projection is also known as *stereographic projection* or *Wulff projection*. The projection of a point A (or B) on the lower reference hemisphere onto the horizontal projection plane is defined as the point where the line OA (or OB) pierces the plane, as shown in Figure 2.11. Here, the point O is the top of the sphere, known as the *zenith*. An area on the lower hemisphere gets distorted when projected onto the plane. An area of $2\pi R^2$ on the surface of the lower hemisphere is projected into an area of πR^2 on the horizontal. The extent of distortion depends on where the area is located. The distortion is more for the regions closer to the projection plane.

2.5.3 Projections of great circles on horizontal planes

A plane with dip ψ and dip direction α can be represented on a lower reference hemisphere, by a great circle or its pole as shown in Figure 2.8. The projections (equal area or equal angle) of this great circle and the pole onto the horizontal reference plane are shown in Figure 2.12. The two features that are important to note are as follows:

1. The larger the dip, the closer is the projection of the great circle to the centre.
2. The larger the dip, the further is the pole from the centre.

These are simple facts you will notice when looking at projections of great circles and poles.

The preceding two points are illustrated in Figure 2.13, which shows the projections of the great circles and poles representing different planes. The concept is the same for equal area or equal angle projections. Now let us see how

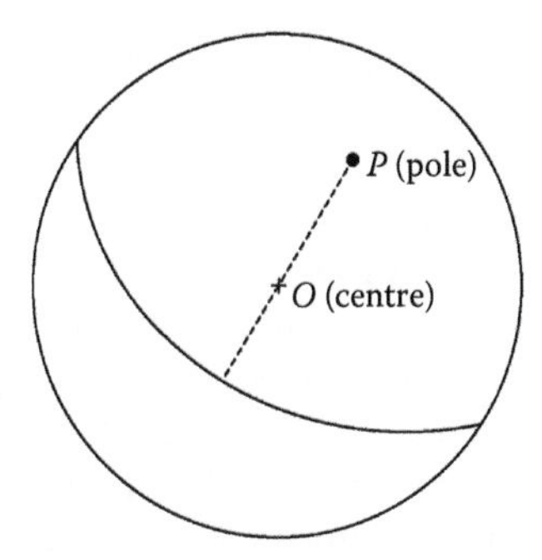

Figure 2.12 Projection of a great circle defining a plane.

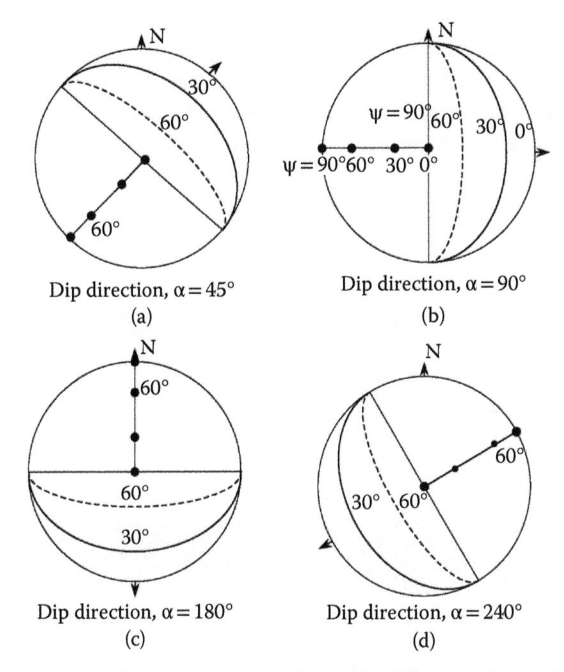

Figure 2.13 Projections of some great circles with different dip and dip directions: (a) $\alpha = 45°$, (b) $\alpha = 90°$, (c) $\alpha = 180°$ and (d) $\alpha = 240°$.

we can use the equatorial and polar stereonets shown in Figure 2.9 (they are in fact equal area stereonets) to precisely define the orientation of a plane.

2.5.4 Polar stereonet

Polar stereonets are used only to plot the pole of a plane, which fully defines the orientation of the plane. A plane dipping towards north will have the pole on a radial line towards south, which corresponds to a dip direction of 0°. The steeper the plane, the further is the pole from the centre. In the polar stereonet, dip (ψ) increases from 0° at the centre to 90° at the perimeter. The concentric circles and the radial lines in Figure 2.14 are at 2° intervals. The dips and the dip directions of the planes represented by poles *A*, *B* and *C* in Figure 2.14 are summarised in Table 2.1. When hundreds of poles representing various discontinuities are plotted, it is possible to identify their concentrations and hence simplify them into a few *sets* of discontinuities that may be easier to analyse. It is recommended to take at least 100 measurements of dip and dip directions in any attempt to identify the orientations of the discontinuities. If necessary, this can be increased further until a clear pattern emerges.

The types of discontinuities plotted in a *pole plot* can be distinguished by using different symbols. Hoek and Bray (1977) suggested using filled circles for faults, open circles for joints and triangles for bedding planes. These days, it is quite common to generate pole plots showing their concentrations and contours using computer programs such as DIPS developed by University of Toronto, now available through Rocscience Inc.

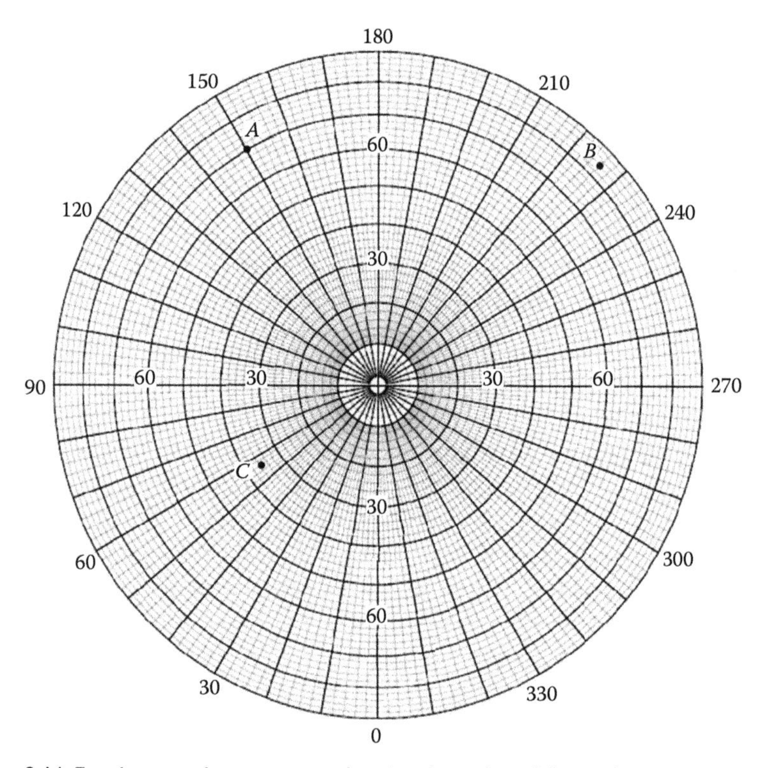

Figure 2.14 Equal area polar stereonet showing the poles of three planes.

Table 2.1 Dips and dip directions of the planes represented by poles *A*, *B* and *C*

Pole	*Dip*, ψ (°)	*Dip direction*, α (°)
A	70	150.5
B	84	226
C	35.5	56

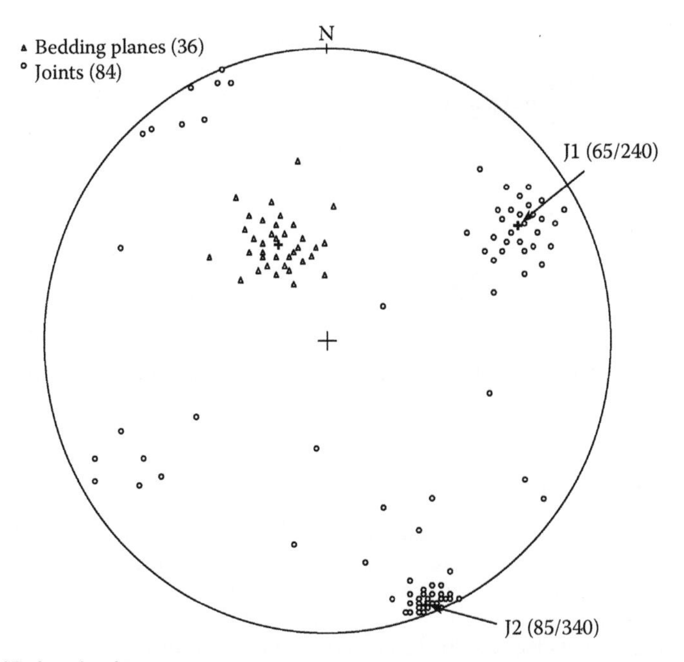

Figure 2.15 A pole plot.

Figure 2.15 shows a pole plot where 36 poles are for the bedding planes and 84 for joints. A close look at the figure clearly shows the pole concentrations and enables one to identify the approximate bedding plane orientation as 30/150 and the orientations of the two joint sets J1 and J2 as 65/240 and 85/340, respectively. The joints in J2 are close to being vertical and hence some of them appear on the opposite side of the pole plot. The thick crosshair shows the average orientation of the bedding plane or joint set.

Figure 2.16 shows the isometric view of the bedding plane and the two other discontinuity planes reflecting the joint sets J1 and J2 in Figure 2.15. The lines of intersection between two planes can be visualised, at least qualitatively, through such isometric views. However, pole diagrams and spherical projections make this work much simpler. When the situation is more complex, it is difficult to draw such isometric views.

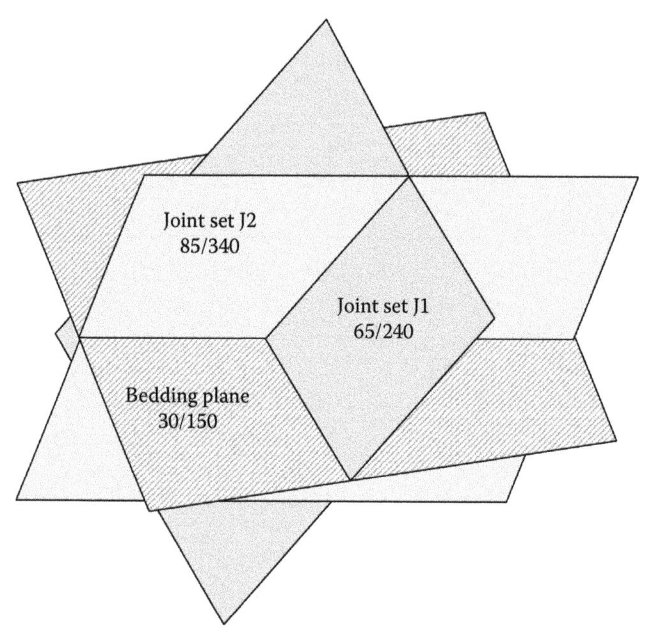

Figure 2.16 Isometric view of the bedding plane (30/150) and joint sets 1 (65/240) and J2 (85/340).

EXAMPLE 2.1

A long cutting is to be made into a hillside with a slope of 70° to the horizontal. The strike of the slope will have an orientation of 30° from north, with the slope falling towards the east. A site investigation exercise at a weathered claystone site produced the following set of measurements for rock bedding and joint orientations.

44/052, 48/052, 48/306, 60/162, 37/130, 52/314, 32/140, 30/290, 26/123, 42/050, 32/130, 52/134, 44/048, 28/126, 68/074, 64/126, 32/124, 48/046, 40/056, 48/300, 46/308, 24/133, 34/120, 60/015, 44/242, 46/308, 52/312, 46/054, 44/208, 44/058 55/306, 46/314, 46/044, 54/305, 46/304, 44/044

Develop a pole plot showing all the above data and identify the number of joint/bedding plane sets. Derive representative orientations for each of the discontinuity sets you have identified.

Solution

The dip and the dip directions of the 36 readings given above are plotted as shown in Figure 2.17.

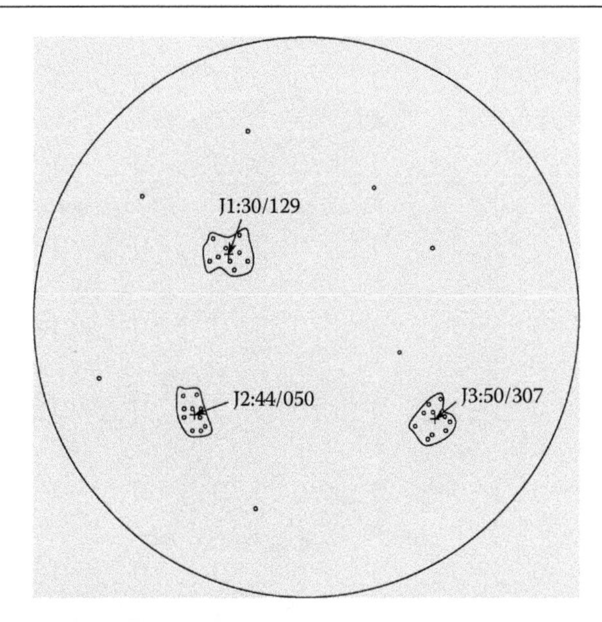

Figure 2.17 Pole plot for Example 2.1.

It is quite clear that there are three sets of discontinuities J1, J2 and J3 with orientations of 30/129, 44/050 and 50/307, respectively. There is some scatter, which we always expect.

2.5.5 Equatorial stereonet

The equatorial projection of the reference sphere shown in Figure 2.9a is reused in Figure 2.18, with some additional labels. Note that we now use this in *plan view* (*not front elevation*), with north, south, east and west directions marked. Also shown in Figure 2.18 are the angles 0–360°, marked along the circumference, reflecting the dip direction of the plane. These are marked along the latitude lines. All meridional lines from north to south reflect the exact locations of the projections of the great circles representing planes dipping at angles of 0–90° towards east or west. We will use these inner meridional lines to precisely draw the projections of great circles of planes dipping at directions that are not necessarily east or west.

Equatorial stereonets can be used to represent planes and their poles. The poles plotted in a polar net fall in the same positions as those plotted on an equatorial net. Therefore, once the pole is marked on a tracing paper placed on top of an equatorial stereonet, it can be verified later by overlaying on a polar stereonet.

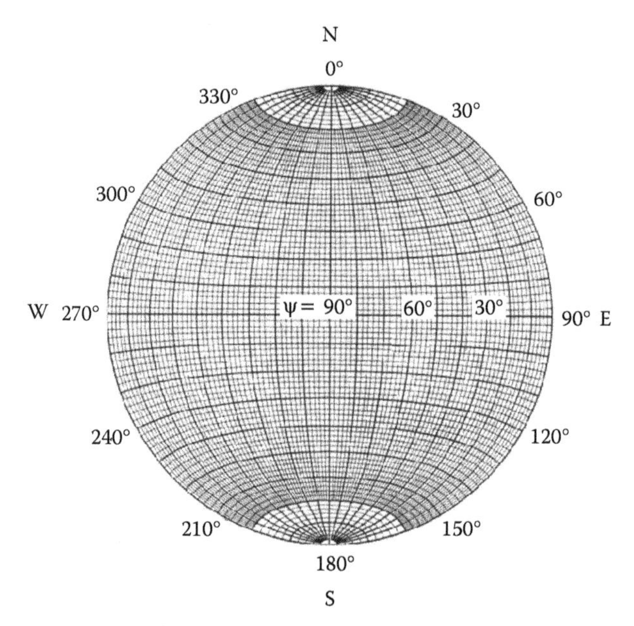

Figure 2.18 Equal area equatorial stereonet.

The great circles representing the planes (i.e., intersection of the plane with the lower reference hemisphere) are projected onto a horizontal plane by the equal area or equal angle projection method. This is best done using a tracing paper, a pin and an equatorial stereonet. This is illustrated here through an example showing a plane with $\psi = 35°$ and $\alpha = 135°$.

Step 1. Place tracing paper over an equatorial net and fix a pin (e.g., thumb tack) at the centre.

Step 2. Trace the circumference of the net and mark north as N on the tracing paper.

Step 3. Count $\alpha = 135°$ along the perimeter and mark the point X on the tracing paper. This is the line of latitude corresponding to 135°.

Step 4. Rotate the tracing paper such that point *X* lies on the E–W axis, *so that we can draw the projection of the great circle (that dips at 35°) precisely.*

Step 5. Trace the meridional circle corresponding to $\psi = 35°$. Mark the pole *P* on the tracing paper, counting $\psi = 35°$ from the centre. Remember, the pole of a horizontal plane is at the centre, and for a vertical plane, it is at the perimeter.

Step 6. Rotate the tracing paper back to its original position such that the 'north mark' N on the tracing coincides with the north on the equatorial stereonet underneath. The great circle and the pole are now in their correct locations (see Figure 2.19).

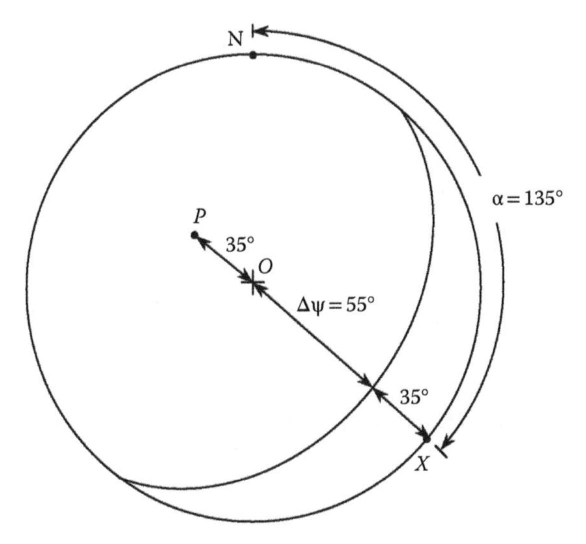

Figure 2.19 Representing a plane (ψ = 35°, α = 135°) using an equatorial stereonet.

Following the preceding steps, it is possible to draw on the same sheet (i.e., tracing paper) any number of great circles, representing different planes. The corresponding poles also can be marked.

A pair of decent stereonets, both polar and equatorial, is the basis for the stereographic projection studies and kinematic analysis discussed in this chapter. These are given in Appendix A.

2.5.6 Intersection of two planes

The discontinuities are often approximated as planes. These planar discontinuities such as faults, joints and bedding planes intersect along straight lines. Now that we have mapped these discontinuities, we should find a way to determine the orientation of the line of intersection between two planes. Figure 2.20a shows two intersecting planes on a lower reference hemisphere. *O* is the centre of the reference hemisphere, which also lies on both planes. The two planes intersect each other along a straight line that meets the reference hemisphere at *X*. Therefore, the radial line *OX* is the line of intersection of the two planes. By mapping the two great circles representing the planes (see Figure 2.20b), their intersection point *X* can be defined. The plunge (ψ) and the trend (α) of the line of intersection *OX* can be determined by following the same steps outlined previously. Rotating Figure 2.20b about the centre O so that *OX* lies on the E–W line enables determination of the plunge ψ of the line of intersection. Extending *OX* to intersect the circumference (i.e., the correct latitude line) enables determination of the trend α, which is measured from north as shown in Figure 2.20b.

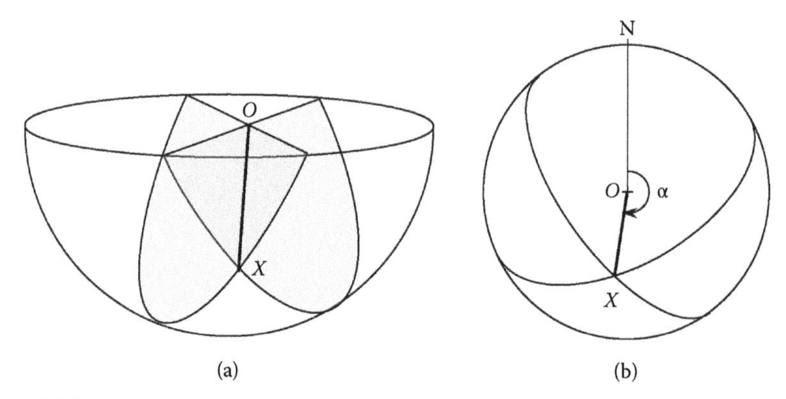

Figure 2.20 Line of intersection between two planes: (a) reference hemisphere and (b) projection.

EXAMPLE 2.2

Find the plunge and trend of the line of intersection between the planes 40/140 and 59/250.

Solution

The steps in the graphical procedure are as follows:

- Place tracing paper on top of an equatorial stereonet and a pin at the centre O.
- Trace the circumference and mark north 'N'. Mark A ($\alpha = 140°$) and B ($\alpha = 250°$) on the perimeter.
- Plot the two great circles following the procedure outlined above and note their intersection point X.
- Rotate the tracing paper until the point X lies on the E–W line on the equatorial stereonet.
- Measure the plunge (ψ) as 33° as shown in Figure 2.21a.
- Rotate the tracing paper to its original position such that the 'N' mark on the tracing paper aligns with the north direction in the equatorial stereonet underneath.
- The line OX defines the trend (direction) of the line and its intersection at the circumference defines the trend as 184° as shown in Figure 2.21b.

2.5.7 Angle between two lines (or planes)

The angle between two planes is the same as the angle between the two radial lines connecting the poles to the centre. In spherical projection, a line is generally represented by its pole, reflecting the plunge and trend. The procedure for measuring the angle between two lines is described through the following example.

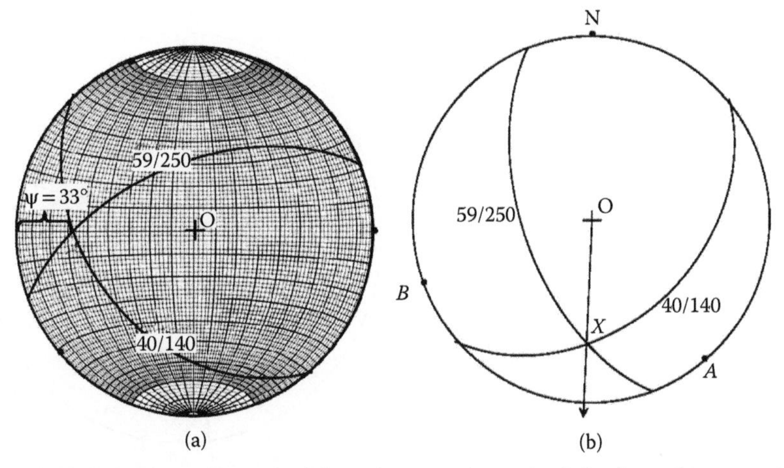

Figure 2.21 Solution to Example 2.2 – plunge and trend of the line of intersection: (a) determining dip and (b) determining dip direction.

EXAMPLE 2.3

Find the angle between two intersecting lines 20/120 and 60/230.

Solution

The steps are outlined as follows:

1. Define the line 20/120 (A) with $\psi = 20°$ and $\alpha = 120°$ following procedure similar to that of locating the pole of a plane (see steps 2 through 6).
2. Place tracing paper on top of an equatorial stereonet and fix a centre pin. Trace the circumference and mark the centre as O and north as N (Figure 2.22a).
3. Locate $\alpha = 120°$ on the perimeter, defining the trend of the above plane. Draw a radial line through this point.
4. Rotate the tracing paper anticlockwise such that the above radial line lies on the E–W line. Count $\psi = 20°$ from the outer circle and mark the point A. Remember, the larger the plunge, the closer the point is to the centre.
5. Rotate the tracing paper to the original position such that the mark 'N' coincides with north on the equatorial stereonet underneath. Now, the point A correctly represents the line 20/120 (see Figure 2.22a).
6. Repeat steps 3–5 for locating the plane 60/230, represented by point B (see Figure 2.22a).
7. Rotate the tracing paper until the two points A and B lie on the *same* meridional great circle (Figure 2.22b). Measure the angle between the two lines by counting the difference in the latitudes of A and B, along the great circle. In this example, A is 69°S and B is 14°N. Therefore, the angle between the two lines OA and OB is 83°.

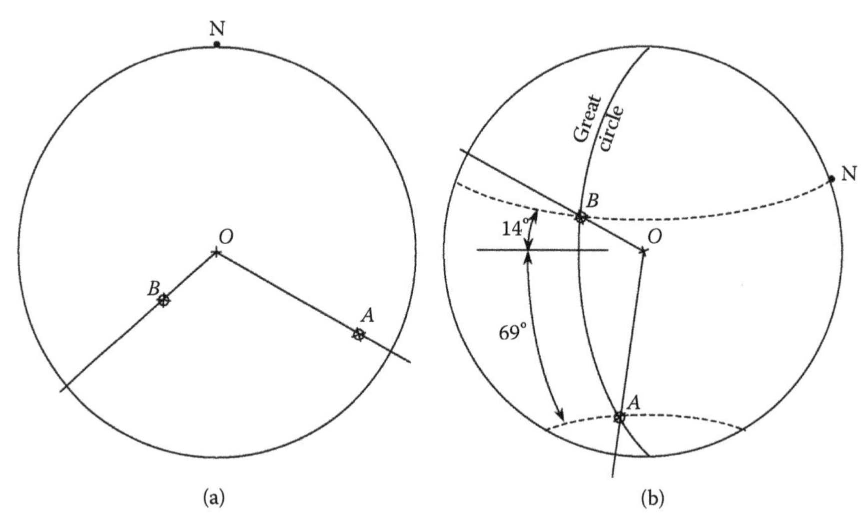

Figure 2.22 Solution to Example 2.3 – angle between two lines: (a) defining the lines and (b) determining angle between the lines.

8. Rotate the tracing paper to original position so that the mark 'N' coincides with north on the equatorial stereonet underneath. The great circle in Figure 2.22b represents the plane that contains the two lines. The dip and the dip direction of this plane can be determined easily.

2.6 SLOPE FAILURE MECHANISMS AND KINEMATIC ANALYSIS

A discontinuity is said to 'daylight' onto the face of the rock slope where the two planes intersect. Figure 2.23 shows a rock slope with three sets of discontinuities *A*, *B* and *C*, shown as dashed lines. Here, the discontinuities *A* and *C* daylight on the slope face. While the discontinuity *A* is of concern due to sliding instability, discontinuity *C* is quite stable. How the discontinuity is oriented has significant influence on the stability.

2.6.1 Slope failure mechanisms

It can be seen that the rock mass above the discontinuity *A* can slide down, leading to *plane failure*, one of the four failure mechanisms suggested by Hoek and Bray (1977). For the plane failure to occur, the dip of the planar discontinuity has to be less than that of the slope face (e.g., $\psi_A < \psi_f$); otherwise, the discontinuity will not daylight on the slope face. In addition, for sliding to take place, the dip directions of both planes must not differ by more than 20° (i.e., $|\alpha_A - \alpha_f| < 20°$) and the dip of the sliding surface has to be greater than the friction angle ϕ.

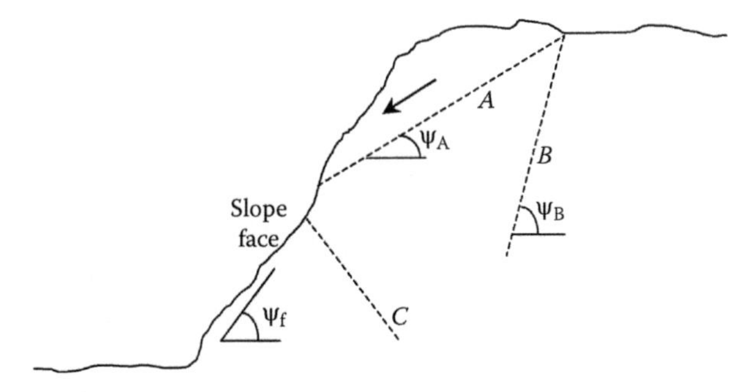

Figure 2.23 Plane failure of a rock slope.

In the case of discontinuity *B*, since the dip ψ_B is greater than the dip of the slope face ψ_f, it will not daylight onto the slope face and therefore plane failure is not possible. Discontinuity *C* does not pose any threat even though it daylights on the slope face.

Two intersecting planes of discontinuities *A* and *B* can daylight on the slope face as shown in Figure 2.24a, where the line of intersection is shown by a dashed line. Here, failure can occur when the wedge enclosed between the two planes slides towards the slope face. This type of failure is known as *wedge failure*, which is one of the four failure mechanisms suggested by Hoek and Bray (1977). Plane failure is a special case of wedge failure where the two planes have the same dip and dip directions.

The spherical representation of the two discontinuities and the slope face is shown in Figure 2.24b. The line of intersection (i) between the two discontinuities defines the direction of sliding, which is shown by the arrow *OX* in Figure 2.24b. The plunge (ψ_i) and the trend (α_i) of this line can be determined as demonstrated earlier. The angle between the two planes of discontinuity can also be determined using the procedure discussed earlier. Generally, larger angles are associated with greater likelihood of wedge failure. In Figure 2.24b, the great circle representing the slope face is shown slightly darker. The arrow *OY* shows the dip direction of the slope face.

For wedge failure to occur, the trend of the line of intersection has to be within 20° from the dip direction of the face of the slope (i.e., $|\alpha_i - \alpha_f| < 20°$). The plunge ($\psi_i$) of this line must be less than the dip (ψ_f) of the slope face so that the line of intersection daylights on the slope face. In addition, the plunge of the line of intersection (ψ_i) has to be greater than the friction angle (ϕ) so that the wedge can slide. These three conditions are the same as the ones for plane failure. The only difference is that here we are looking at the orientation of the line of intersection (along which the slide takes place) rather than a discontinuity plane.

The circular slope failure shown in Figure 2.25 occurs mainly in rock fills, weathered rocks or rocks with closely spaced randomly oriented discontinuities. This three-dimensional slope failure is similar to those occurring

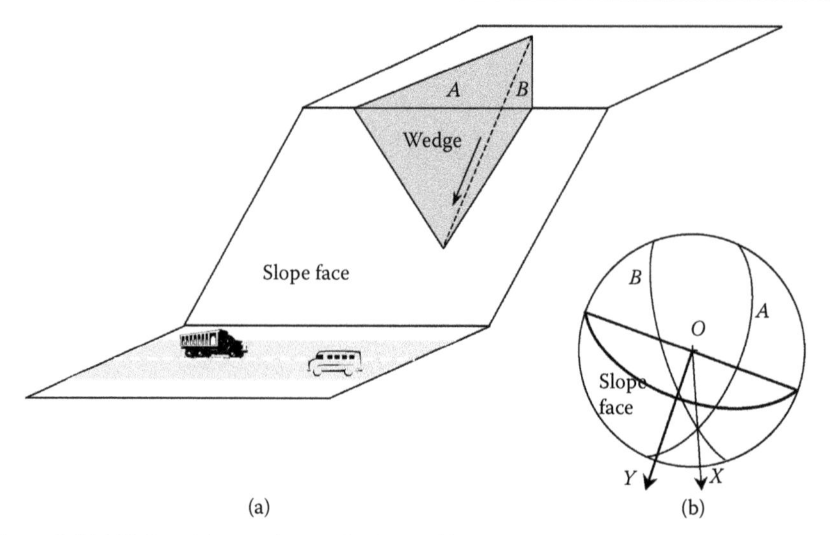

Figure 2.24 Wedge failure of a rock slope: (a) schematic diagram and (b) spherical representation.

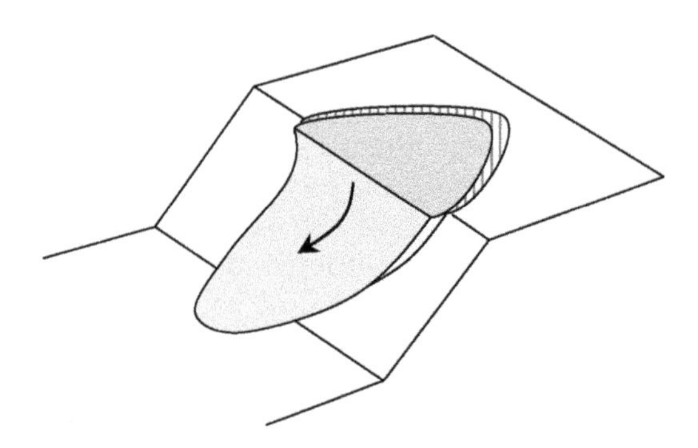

Figure 2.25 Circular failure.

in soils. The fourth failure mechanism suggested by Hoek and Bray (1977) is *toppling failure,* which takes place in hard rocks of columnar structure separated by discontinuities that dip steeply into the slope face.

The spherical projections are valuable tools for identifying the failure mechanisms and carrying out a *kinematic analysis* of the slope stability. Kinematic analysis is a geometric approach that examines the orientations of the discontinuities and the slope face, possible modes of failures and direction of movement in the case of instability. These methods are widely used in structural geology and rock mechanics. While offering a clear picture of the spatial arrangements of the discontinuities, they enable a simple and quantitative analysis of the stability. These are discussed in more detail by Goodman (1989), Hoek and Bray (1977) and Wyllie and Mah (2004).

2.6.2 Kinematic analysis

Figure 2.26a shows the great circle representing the face of a slope, which has a dip of ψ_f and dip direction of $\alpha_f = 270°$, facing west. The dip and the dip direction of a discontinuity *A* are ψ_A and α_A, respectively. For plane failure to occur along the discontinuity *A*, the following conditions must be satisfied:

- $\psi_A < \psi_f$
- $\psi_A > \phi$
- $\alpha_f - 20 \leq \alpha_A \leq \alpha_f + 20$

These three conditions are satisfied only if the dip vector (i.e., line defining the dip and dip direction) of the discontinuity falls within the hatched region in Figure 2.26a. In other words, the lowest point of the great circle representing the discontinuity should lie within the hatched zone for plane failure to occur. For simplicity, the slope is assumed to be facing west. The same procedure applies to slopes facing in any direction. By overlaying tracing paper and rotating about the centre, as in the previous examples, the hatched zone can be defined and the kinematic analysis can be carried out.

Figure 2.26b shows two dashed great circles corresponding to planes of discontinuities *A* and *B*, which intersect at *X*. The line *OX* defines the plunge (ψ_i) and trend (α_i) of the line of intersection between these two planes. Any possible sliding will occur along this line. Wedge failure can occur when all of the following conditions are satisfied:

- $\psi_i < \psi_f$
- $\psi_i > \phi$
- $\alpha_f - 20 \leq \alpha_i \leq \alpha_f + 20$

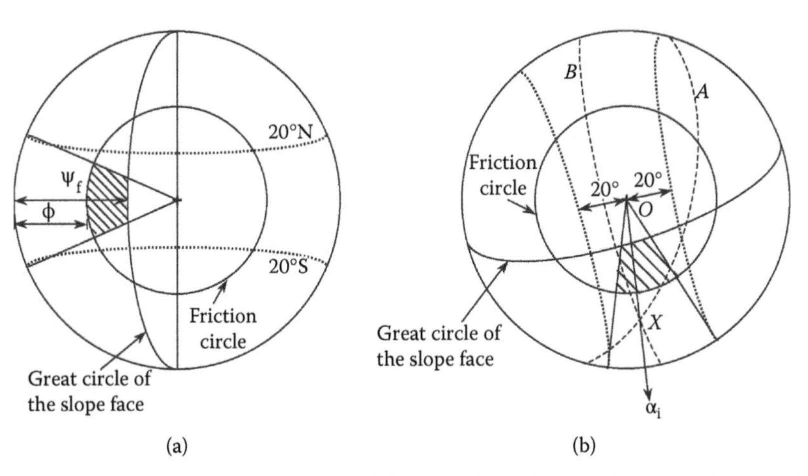

Figure 2.26 Identification of rock slope failure modes: (a) plane failure and (b) wedge failure.

These conditions are satisfied only when X lies within the hatched zone. In the illustration in Figure 2.26b, the second condition is not satisfied (i.e., OX is quite shallow), and hence, wedge failure is unlikely to occur.

EXAMPLE 2.4

Draw the spherical projections of the great circles representing the three joint sets identified in Example 2.1 and the proposed cut on the hillside.

Solution

The great circles of the three joint sets J1 (30/129), J2 (44/050) and J3 (50/307) are shown in Figure 2.27 as continuous lines, along with the cut slope (70/120) shown as a dashed line. The short arrows marked along the perimeter show the dip directions of these four planes.

EXAMPLE 2.5

Check whether planar or wedge failure is likely in Example 2.4. Disregard the friction angle consideration in this exercise.

Solution

Let us check the possibilities for a planar failure. The dotted region in Figure 2.28 shows the region in which the dip vector of the discontinuity should fall for planar failure to occur. The discontinuity set J1 (30/129) certainly appears to have potential to slide and cause a plane failure; the other two joint sets have no possibilities of planar slides. The dip direction of J1 (30/129) and the dip direction of the cut slope (70/120) are too close, within ±20°, and the dip of J1 is less than the dip of the slope. This is recipe for planar failure if the friction angle is less than 30°.

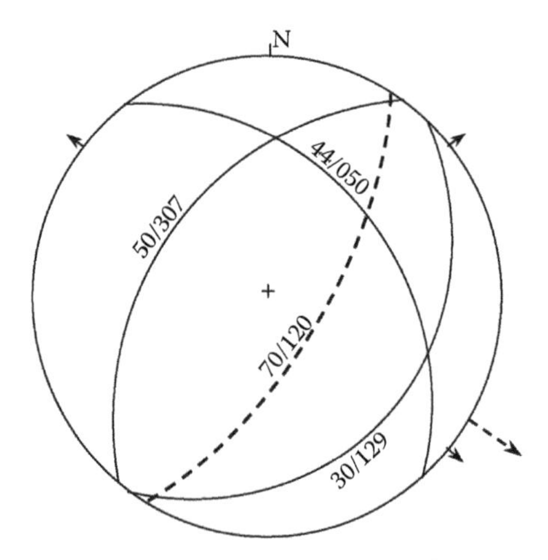

Figure 2.27 Great circles of the three discontinuity sets and the cut slope in Example 2.4.

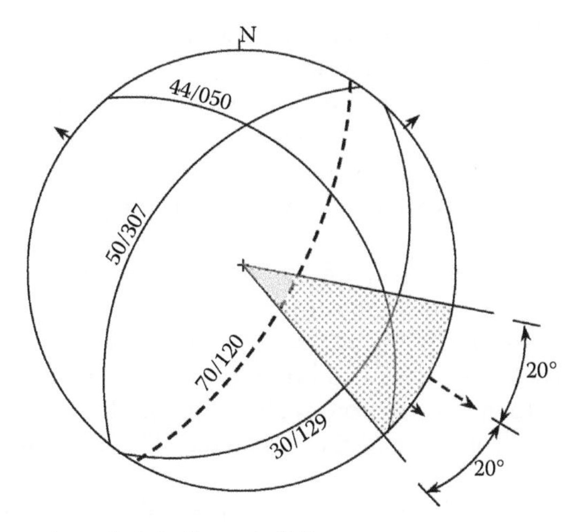

Figure 2.28 Kinematic analysis in Example 2.5.

Let us check the possibilities of wedge failure. The intersection of joint sets J1 (30/129) and J2 (44/050) lies within the dotted region in Figure 2.28. Therefore, there is a possibility of wedge failure. The line of intersection (between J1 and J2) is shallower than the face of the rock slope and hence would daylight on the slope face. The dip direction of this line is within ±20° of the dip direction of the cut slope (70/120).

2.7 SUMMARY

1. Dip and dip direction can fully define the orientation of the plane. Similarly, the orientation of a line is defined by its plunge and trend.
2. It is necessary to be able to quickly visualise the field situation (e.g., discontinuity planes daylighting on slope face) in three dimensions at every occasion.
3. In spherical projection, a line is represented by a point, and a plane is represented by a great circle. They are essentially the intersection of the line and plane, respectively, with the reference hemisphere.
4. In stereographic projection, a plane can be represented by a great circle or pole.
5. There are a few different ways to project the great circle onto a horizontal plane. Equal angle projection and equal area projection are the two common methods. They are different from the traditional projections we use in engineering drawings.
6. There are two separate stereonets we use: equatorial and polar.

7. Poles can be plotted on both equatorial and polar stereonets. The great circles can only be plotted on equatorial stereonets.
8. Poles can be plotted relatively fast on the polar stereonet or on tracing paper placed on top of it without any rotation.
9. Poles marked on an equatorial stereonet can be verified by overlaying the sheet on a polar stereonet. Remember, they fall on the same locations on both nets.
10. The angle between two planes is the same as the angle between the two radial lines connecting their poles.
11. In planar failure, sliding is possible only if the sliding surface daylights on the slope face. In addition, the dip of the sliding plane should be greater than the friction angle, and the dip direction has to be within ±20° from that of the slope face. The same applies to wedge failure as well.
12. In wedge failure, the line of intersection between the two discontinuity planes defines the direction of movement.

Review Exercises

1. List 10 ancient rock-related construction marvels in chronological order, giving the important details about them very briefly.
2. List the different types of discontinuities and emphasise the differences.
3. Carry out a small research on the terms *schistosity, foliation* and *cleavage*, which are different forms of discontinuities, and write a 500-word essay.
4. Four great circles representing planes *A, B, C* and *D* are shown in the following figure. Answer the following.

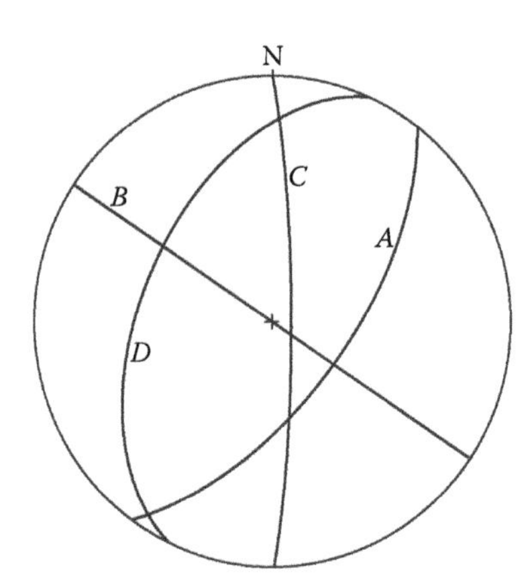

a. Which of the four planes has the largest dip?
   (i) A (ii) B (iii) C (iv) D
b. Which of the four planes has the smallest dip?
   (i) *A* (ii) *B* (iii) *C* (iv) *D*
c. Which of the four planes dips into the northwest quadrant?
   (i) *A* (ii) *B* (iii) *C* (iv) *D*
d. Which of the following is the likely strike direction of plane *A*?
   (i) *N35E* (ii) *S55E* (iii) N35W (iv) S55W
e. Which of the four planes has a dip direction of 295°?
   (i) *A* (ii) *B* (iii) *C* (iv) *D*

5. The poles of four planes *A*, *B*, *C* and *D* are shown in the following figure.

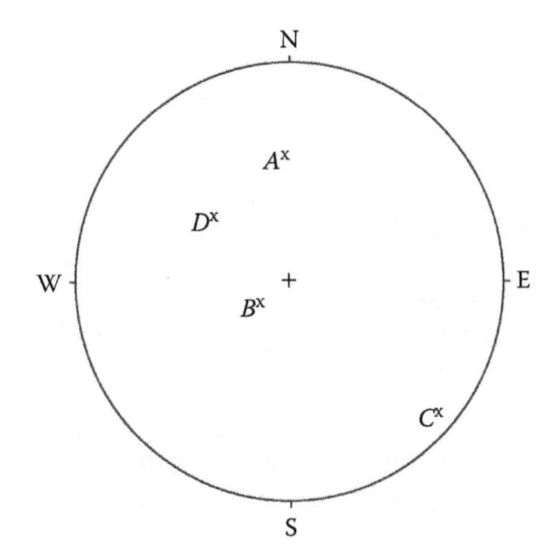

a. Which of the four planes has the largest dip?
   (i) *A* (ii) *B* (iii) *C* (iv) *D*
b. Which of the following is the likely dip direction of plane *A*?
   (i) 0° (ii) 90° (iii) 180° (iv) 270°
c. Which of the four planes has the lowest dip?
   (i) *A* (ii) *B* (iii) *C* (iv) *D*

6. Photocopy the figure from the previous exercise, enlarge it to the size of the stereonets in Appendix A, and draw the great circles of all four planes.
7. State whether the following are true or false.
   a. The apparent dip of a plane can be greater than its true dip.
   b. The pole of a plane plots at the same location in both polar and equatorial stereonets.
   c. The pole is adequate to define a plane.
   d. The diameter of a great circle is the same as the diameter of a reference sphere.
   e. A strike direction of N30E is the same as S30W.

8. The orientations of two joint sets are 50/090 and 60/240. Represent the two on an equatorial stereonet, showing their great circles and poles.
   Using a polar stereonet, check whether the poles you have marked are located at the right places.
   What is the angle of intersection between the two planes?
   What is the orientation of the line of intersection?
   Answer: 77°, 20/163
9. The line of intersection between two joints (i.e., planes) has a plunge of 38° and trend in the northwest quadrant. The first plane has dip and dip direction of 50° and 256°, respectively. What is the trend (i.e., dip direction) of the line of intersection? If the second plane strikes exactly northwest, find its dip and dip direction.
   Answer: 308°; 80°, 225°
10. The line of intersection between two planes has a plunge of 28° and the trend line in the northeast quadrant. One of the two planes has dip and dip direction of 70° and 292°, respectively. What is the trend (i.e., dip direction) of the line of intersection? If the second plane has a strike of 120°, find its dip direction and dip angle.
    Answer: 11°; 30/030
11. During a site investigation at some rock cuts, the following joint orientations were mapped:

25/270	82/230	80/040	90/010	70/140	70/110
80/050	62/110	58/130	90/220	90/035	85/185
85/225	88/025	15/270	75/020	90/200	90/028
80/218	85/210	50/115	90/210	90/045	70/122
30/330	20/260	15/250	88/030	58/105	22/315

    Plot the joint orientations on a polar stereonet.
    How many join sets can you locate? Find the average orientation of each set.
    Answer: Three sets; 20/285, 63/120, 90/210
12. Let us assume you are flying from Townsville, Australia (latitude 18.5° south and longitude 147° east), to either Perth (latitude 31.5° south and longitude 116° east) or Singapore (latitude 1.3° north and longitude 103.8° east). Using the equatorial net, plot the locations of the three cities. Find the distance between (a) Townsville and Perth and (b) Townsville and Singapore. Assume that the radius of the Earth is 6600 km.
    Answer: 3570 km; 5300 km
13. The following question is in several parts which are related.
    a. Using equatorial lower hemisphere projection, represent the two joint orientations J1 (150/40) and J2 (260/50) by their

great circles on tracing paper. In the same plot, show the poles of the two planes as well and give the orientations of the poles.

b. Use a polar stereonet and verify that the poles are plotted at the right place.
c. Determine the orientation of the line of intersection between the two planes.
d. On the tracing paper used for the above exercise, show the line that lies on the plane of joint J1 and is perpendicular to the line of intersection. Repeat this for joint J2. What are the orientations of these two lines? What is the angle between the two lines (and hence the two planes)?
e. Draw the great circle representing the plane normal to the line of intersection between the two planes. What is the orientation of this great circle? Note that the above two lines [from (d)] are lying on this plane.
f. Determine the angle between the two planes by determining the angle between the two radial lines connected to the poles of the planes. Is it the same as in (d)? Discuss.

Answer: 50/330, 40/080; 29/199; 26/094, 37/312, 109°; 61/019; 71°, same as 109°

REFERENCES

Goodman, R.E. (1989). *Introduction to Rock Mechanics*. 2nd edition, Wiley, New York.

Hoek, E. and Bray, E. (1977). *Rock Slope Engineering*. 2nd edition, The Institution of Mining and Metallurgy, London.

Wyllie, D.C. and Mah, C.W. (2004). *Rock Slope Engineering*. 4th edition, Spon Press, London.

Chapter 3

Rock properties and laboratory testing

3.1 INTRODUCTION

Rock mass consists of *intact rock* blocks, separated by various discontinuities that are formed by weathering and other geological processes. Intact rock is an unjointed piece of rock. Rock fragments and rock cores used in laboratory tests are generally all intact rocks. The intact rock itself is a non-homogeneous, anisotropic and inelastic material. The presence of discontinuities in a large scale makes the situation even more complex. The engineering performance of a rock mass under external loadings is very often governed by the strength and orientation of the discontinuities rather than the properties of the intact rock. Other factors that influence rock behaviour are the presence of water and the initial stresses within the rock mass. The discontinuities make the rock mass weaker than the intact rock. In addition, the discontinuities allow access to water, thus compounding the problem. Figure 3.1 shows a relatively steep excavation in a heavily jointed rock.

3.2 ENGINEERING PROPERTIES OF INTACT ROCK

The *unconfined compressive strength,* also known as the uniaxial compressive strength, (UCS) and *Young's modulus* (E) of concrete used in foundations are typically in the order of 30–50 MPa and 25–35 GPa, respectively. The values reported for most intact rocks are significantly greater than the above values. The UCS and E of intact rocks can be on the order on 1–350 MPa and 1–100 GPa, respectively. In the absence of discontinuities, there is very little need for us to worry about the adequacy of the intact rock as support for most foundations as these values are quite high. However, the presence of discontinuities can make a big difference and make one feel that the parameters of intact rock are irrelevant. In other words, the discontinuities will have a much greater bearing on the way the rock mass behaves under the applied loadings.

Figure 3.1 Rock mass with several discontinuities.

This section discusses the techniques adopted in the field for obtaining intact rock specimens and those adopted in the laboratory for preparing the specimens for specific tests. The different standards available for the laboratory testing of rocks are also briefly discussed.

3.2.1 Rotary versus percussion drilling

Rotary drilling and percussion drilling are two different ways of drilling into the rock overburden. In percussion drilling, the drill bit is repeatedly hammered into the rock. In rotary drilling, a sharp rotating drill bit is advanced into the ground, exerting a downward pressure as well. To obtain good quality rock cores for laboratory tests, rotary drilling is a better option and is more common.

3.2.2 Rock coring

Rock specimens from the ground are recovered through *coring*, a procedure different from sampling in soils. The high strength of the rock makes it necessary to use thick-walled core barrels (tubes or pipes) with tips made of some of the hardest materials such as diamond or tungsten carbide. The rotary drill grinds away an annular zone around the specimen and advances into the ground while the cuttings are washed out by circulating water, in a manner similar to wash borings in soils. The central rock core is collected within the core barrel, which can typically retain cores of 0.5–3.0 m in length. The coring process subjects the cores to some torsion and significant mechanical disturbance. In addition, the core can undergo swelling and get contaminated by the drilling fluid, especially if the rock is weak or heavily fractured. These disturbances can be minimised by using *double-tube* or *triple-tube core barrels*. The cores collected are placed in sequence in a core box (Figure 3.2), with

(a)

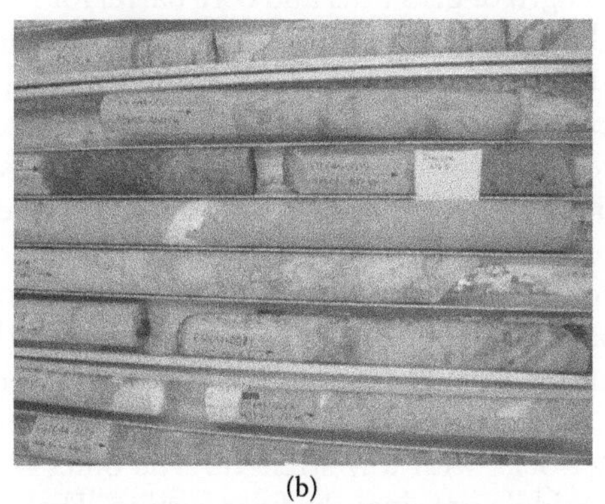

(b)

Figure 3.2 Intact rock cores received at James Cool University laboratory: (a) several core boxes from a large project and (b) core box.

the borehole number and depth marked, for transporting to the laboratory for further testing and analysis. They also provide a tangible and accurate representation of the underlying rock formations.

The drill rod, core barrel and casing are slightly different in diameter. The early drill holes had diameters of 38 mm (1½ in.), 51 mm (2 in.), 63.5 mm (2½ in.) and 76 mm (3 in.), matching the standard steel pipes available and they were given designations of E, A, B and N, respectively. With some standardisation worldwide in 1930, an 'X' was added. H and P are larger sizes that were introduced later.

Some of the common core sizes and their standard designations are given in Table 3.1. The first letter of the symbol (e.g., E, A, B, N, H and P) identifies

Table 3.1 Core size designations and nominal diameters

	Nominal core diameter		Nominal hole diameter	
Symbol	*(mm)*	*(in.)*	*(mm)*	*(in.)*
AQ	27.0	$1\frac{1}{16}$	48.0	$1\frac{57}{64}$
BQ	36.5	$1\frac{7}{16}$	60.0	$2\frac{23}{64}$
NQ	47.6	$1\frac{7}{8}$	75.8	$2\frac{63}{64}$
HQ	63.5	$2\frac{1}{2}$	96.0	$3\frac{25}{32}$
PQ	85.0	$3\frac{11}{32}$	122.6	$4\frac{53}{64}$
EX	22.2	$\frac{7}{8}$	36.5	$1\frac{7}{16}$
AX	30.2	$1\frac{3}{16}$	47.6	$1\frac{7}{8}$
BX	41.3	$1\frac{5}{8}$	58.7	$2\frac{5}{16}$
NX	54.0	$2\frac{1}{8}$	74.6	$2\frac{15}{16}$

the core diameter. The second letter Q signifies *wire line drilling*, a technique widely used for deep drilling to minimise the time lost in removing and reinserting the entire length of drill rods and core barrel for recovering the cores. Instead, the core barrel is lowered down a wire line inside the outer barrel, which extends to the full depth of the hole. Upon reaching the bottom of the hole, the core barrel is latched inside the outer barrel and drilling proceeds.

Single-tube core barrels are the most rugged and least expensive. They are used in the beginning of the drilling operation and are adequate in homogeneous hard intact rock mass or in situations when very good quality sampling is not required. Double-tube core barrels are the most common and are often used with NX cores. While the outer barrel moves with the cutting bit, the inner barrel retains the core. In fractured or highly weathered rocks, to minimise the disturbance, triple-tube core barrels are preferred. They are also effective on brittle rocks with low strength. The outer barrel does the first cutting, while the middle one does the finer cutting. The third and the innermost barrel retains the core. This process reduces the heat generated at the cutting edge that can otherwise damage the core. A '3' or 'TT' is added to the two-letter symbol given in Table 3.1 for triple-tube core barrels (i.e., PQ3).

3.2.3 Rock quality designation

When attempting to obtain a rock core over a certain depth, due to the presence of joints and fractures, a significant length may be 'lost'. This can be seen as a measure of the quality of the intact rock. Two similar parameters commonly used to ascertain the quality of intact rock based on the drill record are *core recovery ratio* (CR) and *rock quality designation* (RQD). Core recovery ratio is defined as

$$\text{CR}\ (\%) = \frac{\text{Length of rock core recovered}}{\text{Total length of the core run}} \times 100 \tag{3.1}$$

Rock quality designation (RQD) is a modified measure of core recovery, defined as (Deere, 1964)

$$\text{RQD (\%)} = \frac{\sum \text{Lengths of core pieces equal to or longer than 100 mm}}{\text{Total length of the core run}} \times 100 \tag{3.2}$$

The RQD is a simple and inexpensive way to recognise low-quality rock zones that may require further investigation. The RQD, corresponding descriptions of in situ rock quality, and the allowable foundation bearing pressures as given by Peck et al. (1974) are summarised in Table 3.2. The lengths are measured along the centre line of the core. In computing the RQD, breaks that are caused by the drilling process are ignored. RQD is a parameter used in some of the popular rock mass classification systems discussed in Chapter 4.

RQD and CR are influenced by the drilling technique and the size of the core barrel. The International Society for Rock Mechanics (ISRM) recommends RQD be computed from double-tube NX cores of 54 mm diameter. However, ASTM D6032 permits core diameters from 36.5 mm (BQ) to 85 mm (PQ) to be used for RQD computations, while suggesting NX (54 mm) and NQ (47.6 mm) as the optimal core diameters for this purpose.

The cores recovered from the ground are tested in the laboratory to determine strength and deformation characteristics, durability and hardness. Some of the common laboratory tests on rocks are as follows:

- Uniaxial compressive strength test
- Brazilian indirect tensile strength test
- Point load strength index test
- Schmidt hammer test
- Slake durability test
- Triaxial test

Table 3.2 RQD, in situ rock quality description, and allowable bearing pressure

RQD (%)	*Rock quality*	*Allowable bearing pressure (MPa)*
0–25	Very poor	1–3
25–50	Poor	3–6.5
50–75	Fair	6.5–12
75–90	Good	12–20
90–100	Excellent	20–30

Source: Peck, R.B. et al., *Foundation Engineering*, John Wiley & Sons, New York, 1974.

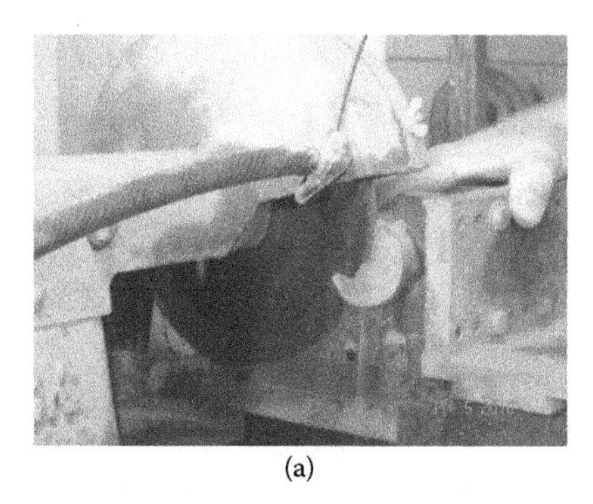

(a)

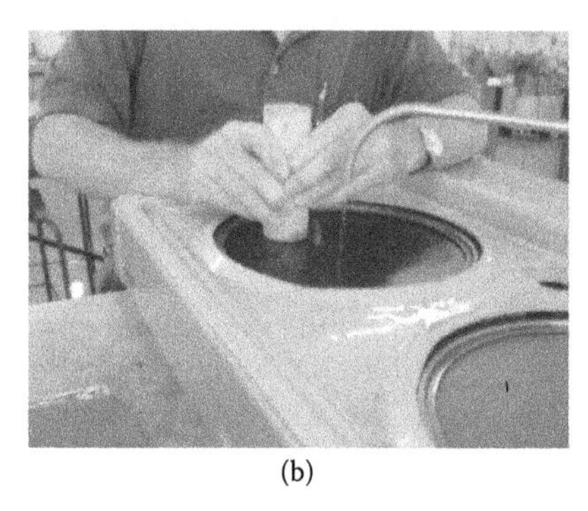

(b)

Figure 3.3 Specimen preparation: (a) cutting the ends using a diamond saw and (b) polishing the ends.

3.2.4 Specimen preparation

Laboratory tests such as UCS, triaxial and point load tests require good quality cylindrical specimens that have their ends cut parallel and flat, such that they are perpendicular to the longitudinal axis of the specimen. The standard requirements are discussed in ASTM D4543. Figure 3.3a shows a specimen being cut by a diamond saw. Then the ends are further smoothened using a surface grinder and polished (Figure 3.3b) to minimise friction during loading. Non-parallel ends can induce eccentricity in the applied loads. Roughness at the ends can mean that the applied stresses are no longer principal stresses. Applying capping materials (e.g., sulphur) to the ends is not generally recommended with rock specimens.

3.2.5 Standards

Similar to the International Society of Soil Mechanics and Geotechnical Engineering (ISSMGE) that looks after the research and professional practice in soil mechanics and geotechnical engineering, there is a society for rock mechanics too. The ISRM is a non-profit scientific organisation that has more than 5000 members representing 46 national groups (http://www.isrm.net). It was founded in 1962 at Karlsruhe University by Professor Leopold Mueller. It appointed the Commission on Standardization of Laboratory and Field Tests on Rock in 1967, which later became The Commission on Testing Methods. The commission proposed 'Suggested Methods' for various rock tests that have been adopted worldwide and were published from time to time in the *International Journal of Rock Mechanics and Mining Sciences & Geomechanics Abstracts*, Pergamon Press, United Kingdom. These were compiled by Professor Ted Brown (1981) of University of Queensland, Australia, as the ISRM 'Yellow Book'. This was later updated by Professor Ulusay of Hacettepe University, Turkey and Professor Hudson, formerly of Imperial College, United Kingdom, in 2007 as the 'Blue Book', which is a one-stop shop for all relevant ISRM-suggested methods for rock testing. The test procedures for rocks, described in this chapter, are mainly based on the ISRM-suggested methods, with references to ASTM (American Society for Testing Materials) and Australian Standards as appropriate. The United States, United Kingdom, Canada, South Africa and Australia are some of the countries that had pioneering roles in the developments in rock mechanics, including the laboratory test methods.

3.3 UNIAXIAL COMPRESSIVE STRENGTH TEST

The UCS test is also known as the *uniaxial compressive strength* test. Here, a cylindrical rock specimen is subjected to an axial load, without any lateral confinement. The axial load is increased gradually until the specimen fails. The vertical normal stress on the specimen, when failure occurs, is known as the unconfined compressive strength or uniaxial compressive strength, fondly known as UCS. By monitoring the vertical deformations, the vertical normal strains can be computed. By plotting the stress–strain curve, the Young's modulus (E) can be determined. By monitoring the lateral or circumferential deformation, the Poisson's ratio can be computed too.

3.3.1 Soils versus rocks

What is the dividing line between hard soil and soft rock? When do we call a material a rock rather than a soil? A commonly used but rather arbitrary cut-off is the uniaxial compressive strength of 1 MPa. Soils have their UCS and E generally quoted in kPa and MPa, respectively. In rocks, they are orders of magnitude greater and are given in MPa and GPa respectively.

Saturated clays under undrained conditions are generally analysed using the total stress parameters c_u and ϕ_u. Here, c_u is the undrained shear strength and ϕ_u is the friction angle in terms of total stresses. In saturated clays, the

Mohr–Coulomb failure envelope in terms of total stresses is horizontal and hence $\phi_u = 0$. The unconfined compression test is one of the many ways of deriving the undrained shear strength of a clay. The UCS of a clay, denoted often by q_u in geotechnical literature, is twice the undrained shear strength c_u when $\phi_u = 0$.

The same principle holds in rocks too. Uniaxial compressive strength often denoted by σ_c in rock mechanics literature, is the most used rock strength parameter in rock mass classification and rock engineering designs. Unlike saturated undrained clays, the friction angle of a rock specimen is not zero, and hence, the Mohr–Coulomb failure envelope is not horizontal. It can be shown from the Mohr circle that

$$\sigma_c = \frac{2c\cos\phi}{1-\sin\phi} \tag{3.3}$$

where c and ϕ are the cohesion and friction angles of the rock, respectively.

3.3.2 Test procedure

The test is quite simple and the interpretation is fairly straightforward. A cylindrical core of at least 54 mm in diameter (NX core) and length/diameter ratio of 2.0–3.0 (ISRM suggests 2.5–3.0 and ASTM D 7012 suggests 2.0–2.5) is subjected to an axial load that is increased to failure. The specimen is loaded axially using spherical seating, at a constant rate of strain or stress such that it fails in 5–15 minutes. Alternatively, the stress rate shall be in the range of 0.5–1.0 MPa/s. The axial loads at failure can be very large for large diameter cores in good quality intact igneous rocks. Uniaxial compressive strength is the maximum load carried by the specimen divided by the cross-sectional area.

The change in the specimen length is measured throughout the test, using a dial gauge or an LVDT (linear variable differential transformer). These days, it is common to use sophisticated data acquisition systems that would keep track of the load–deformation data. Figure 3.4a shows a UCS test in progress on an MTS universal testing machine with axial load capacity of 1000 kN and a data acquisition system. To prevent injury from flying rock fragments on failure, a protective shield should be placed around the test specimen as shown in the figure. The load–displacement plot generated from the MTS machine, for a rock specimen, is shown in Figure 3.4b.

From the load and displacement measured throughout the loading, the stress–strain plot can be generated. From the stress–strain plot, Young's modulus (E) can be computed. Young's modulus is a measure of the rock stiffness, which is required in modelling the rock and for computing deformations, where the rock is assumed to be an elastic continuum. You may recall *Hooke's law* from the study of the strength of materials, which states

(a)

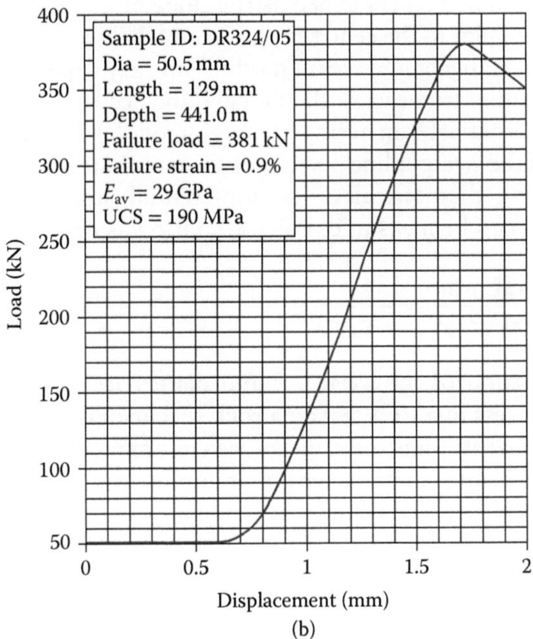

(b)

Figure 3.4 (a) UCS test on an MTS universal testing machine and (b) load–displacement plot.

that stress is proportional to the strain in a *linear elastic* material. Young's modulus is the slope of the stress–strain plot. In reality, rocks are not linearly elastic and the stress–strain plot is not a straight line. There are a few different ways of defining Young's modulus here. The *tangent modulus* (E_t) is defined as the slope of a tangent to the stress–strain plot (Figure 3.5a).

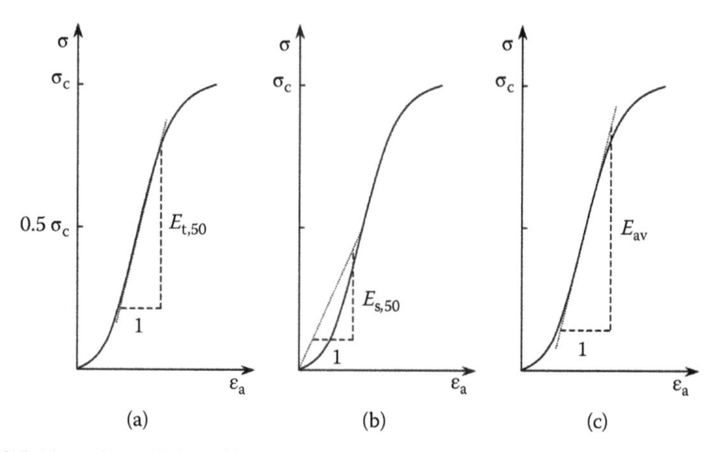

Figure 3.5 Young's modulus: (a) tangent modulus, (b) secant modulus and (c) average modulus.

The *secant modulus* (E_s) is defined as the slope of a line joining a point on the stress–strain plot to the centre (Figure 3.5b). When the stress–strain plot is not linear, the tangent and secant moduli can vary depending on the stress level. It is common to measure the tangent and secant Young's modulus at 50% of σ_c. Alternatively, an average Young's modulus E_{av} can be determined as the slope of the straight line portion of the stress–strain plot (Figure 3.5c).

By measuring diametrical or circumferential strains during loading, Poisson's ratio can be measured. Poisson's ratio v is defined as

$$\nu = -\frac{\text{Lateral strain}}{\text{Axial strain}} = -\frac{\varepsilon_d}{\varepsilon_a} \tag{3.4}$$

Typical variation of the axial (ε_a) and diametrical (ε_d) strains with the applied axial stress in a UCS test on a rock specimen is shown in Figure 3.6. Here, diametrical strain is the same as the circumferential strain, defined as the ratio of the change in diameter (or circumference) to the original diameter (or circumference). The volumetric strain (ε_{vol}) of the specimen is given by

$$\varepsilon_{vol} = \varepsilon_a + 2_d \tag{3.5}$$

Poisson's ratio for a common engineering material varies in the range of 0–0.5. Typical values of Poisson's ratio for common rock types are given in Table 3.3. Hawkes and Mellor (1970) discussed various aspects of the UCS laboratory test procedure in great detail. Typical values of the uniaxial compressive strength for some major rock types, as suggested by Hudson (1989), are given in Figure 3.7. As seen here, the UCS values are in the range of 0–350 MPa for most rocks. The axial strain at failure is a measure of the ductility of the intact rock. Qualitative descriptions of materials as ductile,

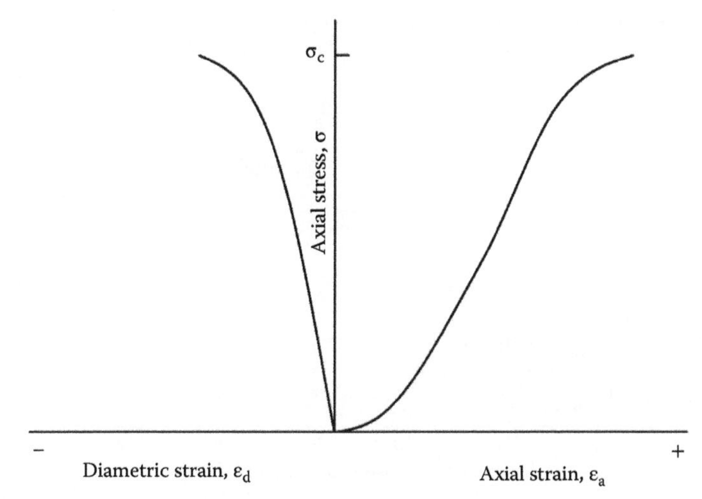

Figure 3.6 Variation of axial and diametrical strains with the applied axial stress.

Table 3.3 Typical values of Poisson's ratio for rocks

Rock type	ν
Andesite	0.20–0.35
Basalt	0.10–0.35
Conglomerate	0.10–0.40
Diabase	0.10–0.28
Diorite	0.20–0.30
Dolerite	0.15–0.35
Dolomite	0.10–0.35
Gneiss	0.10–0.30
Granite	0.10–0.33
Granodiorite	0.15–0.25
Greywacke	0.08–0.23
Limestone	0.10–0.33
Marble	0.15–0.30
Marl	0.13–0.33
Norite	0.20–0.25
Quartzite	0.10–0.33
Rock salt	0.05–0.30
Sandstone	0.05–0.40
Shale	0.05–0.32
Siltstone	0.05–0.35
Tuff	0.10–0.28

Source: Gercek, H., *Int. J. Rock Mech. Min. Sci.*, 44, 1–13, 2007.

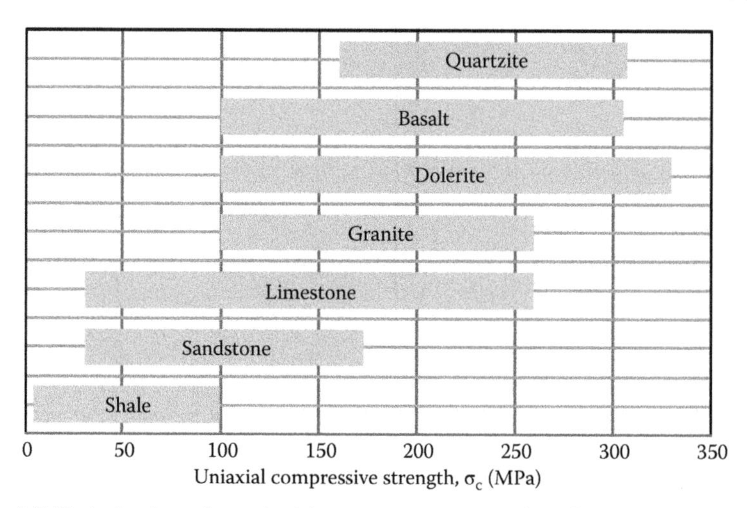

Figure 3.7 Typical values for uniaxial compressive strengths of common rock types. (Adapted from Hudson, J.A., *Rock Mechanics Principles in Engineering Practice*, Butterworths, London, 1989.)

Table 3.4 Relative ductility based on axial strain at peak load

Classification	*Axial strain (%)*
Very brittle	<1
Brittle	1–5
Moderately brittle[a] (transitional)	2–8
Moderately ductile	5–10
Ductile	> 10

Source: Handin, J., *Handbook of Physical Contacts*, Geological Society of America, New York, 1966.

[a] Note the overlap.

brittle and so on based on failure strains, as suggested by Handin (1966), are given in Table 3.4.

Young's modulus and Poisson's ratio are the two crucial parameters in defining the rock behaviour when it is assumed to behave as a linear elastic material, obeying Hooke's law. They are related to the bulk modulus K and shear modulus G by

$$K = \frac{E}{3(1-2\nu)} \tag{3.6}$$

and

$$G = \frac{E}{2(1+\nu)} \tag{3.7}$$

EXAMPLE 3.1

A 50.5-mm-diameter, 129-mm-long rock specimen is subjected to a uniaxial compression test. The load–displacement plot is shown in Figure 3.4b. Determine the uniaxial strength and Young's modulus of the intact rock specimen.

Solution

Noting that there was no load for displacement up to 0.6 mm, the origin (i.e., the load axis) is shifted to displacement of 0.6 mm. The cross-sectional area A of the specimen is given by

$$A = \pi \times 25.25^2 = 2003.0 \text{ mm}^2$$

The failure load = 381 kN

∴ UCS = 381,000/2003 MPa = 190.2 MPa

Considering the linear segment of the load–displacement plot between displacements of 1.0 and 1.5 mm in Figure 3.4b,

$$E = \frac{\Delta P \times L}{\Delta L \times A} = \frac{225{,}000 \times 129}{0.5 \times 2003} = 28{,}982 \text{ MPa} = 29.0 \text{ GPa}$$

A semi-quantitative classification of rocks, based on the uniaxial compressive strength and Young's modulus, proposed by Hawkes and Mellor (1970), is shown in Figure 3.8. Here, the *modulus ratio* is the ratio of the Young's modulus E to the uniaxial compressive strength σ_c. In concrete, this ratio is about 1000, which is well above the upper end of the values for rocks. The cut-off values used for the UCS in Figure 3.8 were later revised by ISRM (1978c), which are discussed later in Chapter 4 (see Table 4.1). Typical values of modulus ratios of various rock types, suggested by Hoek and Diederichs (2006), are summarised in Table 3.5.

In clays, the ratio of undrained Young's modulus to the undrained shear strength is expressed as a function of the overconsolidation ratio and the plasticity index, and this varies in the range of 100–1500. Note that undrained shear strength is half of UCS. Therefore, similar modular ratios for clays are in the range of 50–750.

Generally, there is significant reduction in the uniaxial compressive strength with increasing specimen size, as evident from Figure 3.9 (Hoek and Brown, 1980). The uniaxial compressive strength of a d-diameter specimen $\sigma_{c,d}$ and a 50-mm-diameter specimen $\sigma_{d,50}$ are related by

$$\sigma_{c,50} = \sigma_{c,d}\left(\frac{d}{50}\right)^{0.18} \tag{3.8}$$

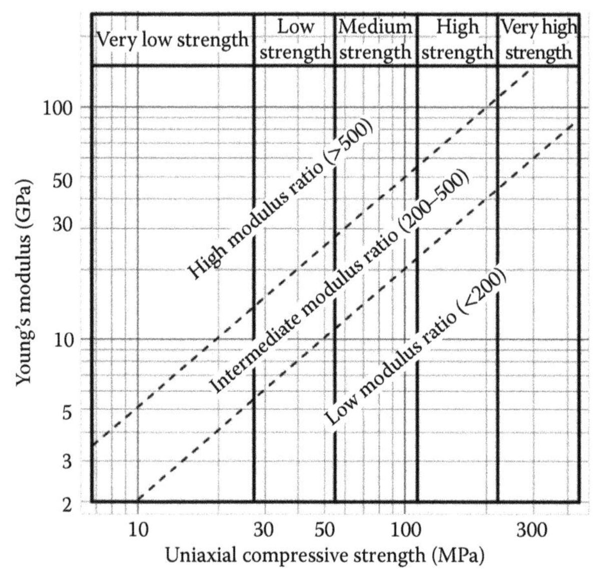

Figure 3.8 Rock classification based on UCS and Young's modulus. (Adapted from Hawkes, I. and M. Mellor, *Eng. Geol.*, 4, 179–285, 1970 and Deere, D.U. and R.P. Miller, Engineering classification and index properties of intact rock. *Report AFWL-TR-65-116, Air Force Weapon Laboratory (WLDC)*, Kirtland Airforce Base, New Mexico, 1966.)

Table 3.5 Typical values of modulus ratios

	Texture			
	Coarse	*Medium*	*Fine*	*Very fine*
Sedimentary	Conglomerates 300–400	Sandstones 200–350	Siltstones 350–400	Claystones 200–300
	Breccias 230–350		Greywackes 350	Shales 150–250[a]
				Marls 150–200
	Crystalline limestone 400–600	Sparitic limestone 600–800	Micritic limestone 800–1000	Dolomite 350–500
		Gypsum (350)[c]	Anhydrite (350)[c]	Chalk 1000[b]
Metamorphic	Marble 700–1000	Hornfels 400–700	Quartzite 300–450	
		Metasandstone 200–300		
	Migmatite 350–400	Amphibolites 400–500	Gneiss 300–750[a]	
		Schists 250–1100[a]	Phyllites/mica schist 300–800[a]	Slates 400–600[a]

Table 3.5 Typical values of modulus ratios (*Continued*)

	Texture			
	Coarse	*Medium*	*Fine*	*Very fine*
Igneous	Granite[b] 300–550	Diorite[b] 300–350		
	Granodiorite 400–450			
	Gabro 400–500	Dolerite 300–400		
	Norite 350–400			
	Porphyries (400)[c]		Diabase 300–350	Peridotite 250–300
		Rhyolite 300–500	Dacite 350–450	
		Andesite 300–500	Basalt 250–450	
	Agglomerate 400–600	Volcanic breccia (500)[c]	Tuff 200–400	

Source: Hoek, E. and M.S. Diederichs, *Int. J. Rock Mech. Min. Sci.*, 43, 203–215, 2006.

[a] Highly anisotropic rocks: the modulus ratio will be significantly different if normal strain and/or loading occurs parallel (high modulus ratio) or perpendicular (low modulus ratio) to a weakness plane. Uniaxial test loading direction should be equivalent to field application.

[b] Felsic granitoids: coarse-grained or altered (high modulus ratio), fine-grained (low modulus ratio).

[c] No data available; estimated on the basis of geological logic.

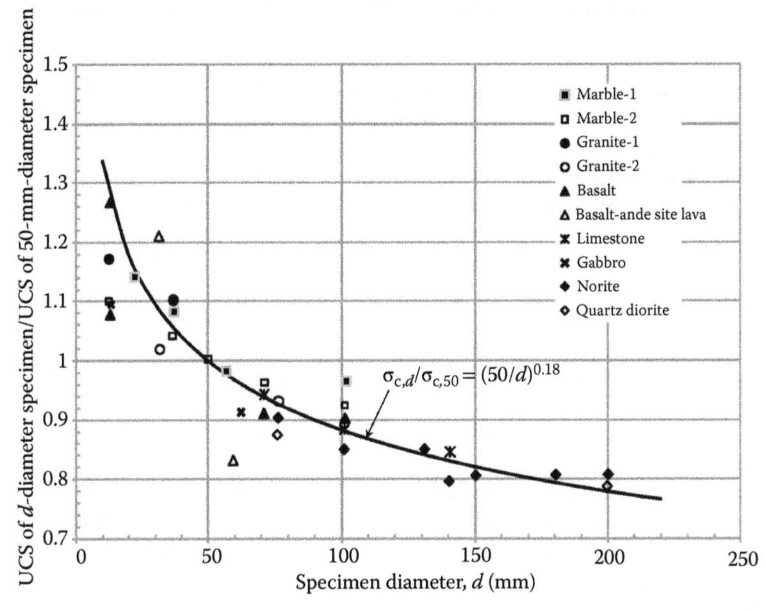

Figure 3.9 Influence of specimen size on UCS. (After Hoek, E. and E.T. Brown, *Underground Excavations in Rock*, Institution of Mining and Metallurgy, London, 1980.)

The reduction is probably due to the fact that the larger specimens include more grains, thus enabling greater tendency to fail around these grain surfaces.

Some typical values of the uniaxial compressive strength, Young's modulus, modulus ratio and Poisson's ratio are given in Table 3.6 (Goodman, 1980). It may be useful to cross-check your laboratory data against these values.

Table 3.6 Typical values of σ_c, E, modulus ratio and ν

Rock description	σ_c *(MPa)*	E *(GPa)*	E/σ_c	ν
Fine-grained slightly porous Berea sandstone	73.8	19.3	261	0.38
Fine- to medium-grained friable Navajo sandstone	214.0	39.2	183	0.46
Calcite cemented medium-grained Tensleep sandstone	72.4	19.1	264	0.11
Argillaceous Hackensack siltstone cemented with hematite	122.7	26.3	214	0.22
Monticello Dam greywacke – Cretaceous sandstone	79.3	20.1	253	0.08
Very fine crystalline limestone from Solenhofen, Bavaria	245.0	63.7	260	0.29
Slightly porous, oolitic, bioclastic limestone, Bedford, Indiana	51.0	28.5	559	0.29
Fine-grained cemented and interlocked crystalline Tavernalle limestone	97.9	55.8	570	0.30
Fine-grained Oneota dolomite with interlocking granular texture	86.9	43.9	505	0.34
Very fine-grained Lockport dolomite, cemented granular texture	90.3	51.0	565	0.34
Flaming Gorge shale, Utah	35.2	5.5	157	0.25
Micaceous shale with kaolinite clay mineral, Ohio	75.2	11.1	148	0.29
Dworshak dam granodiorite gneiss, fine-medium grained, with foliation	162.0	53.6	331	0.34
Quartz mica schist ⊥ schistosity	55.2	20.7	375	0.31
Fine-grained brittle massive Baraboo quartzite, Wisconsin	320.0	88.3	276	0.11
Uniform fine-grained massive Taconic white marble with sugary texture	62.0	47.9	773	0.40
Medium-coarse grained massive Cherokee marble	66.9	55.8	834	0.25
Coarse-grained granodiorite granite, Nevada	141.1	73.8	523	0.22
Fine-medium grained dense Pikes Peak granite, Colorado	226.0	70.5	312	0.18
Cedar City tonalite, Utah – somewhat weathered quartz monzonite	101.5	19.2	189	0.17
Medium-grained Palisades diabase, New York	241.0	81.7	339	0.28
Fine olivine basalt, Nevada	148.0	34.9	236	0.32
John Day basalt, Arlington, Oregon	355.0	83.8	236	0.29
Nevada tuff – welded volcanic ash, with 19.8% porosity	11.3	3.6	323	0.29

Source: Goodman, R.E., *Introduction to Rock Mechanics*, John Wiley & Sons, New York, 1980.

3.4 INDIRECT TENSILE STRENGTH TEST

Unlike soils, rocks can carry some tensile stresses. The tensile strength of a rock is required in most designs, analysis and numerical modelling of excavation, tunnelling, slope stability and so on. On rock specimens, it is difficult to carry out a direct tensile strength test in the same way we test steel specimens. The main difficulties are (1) in gripping the specimens without damaging them and applying stress concentrations at the loading grip and (2) in applying the axial load without eccentricity. The *indirect tensile strength test*, also known as the *Brazilian test*, is an indirect way of measuring the tensile strength of a cylindrical rock specimen having the shape of a disc. The rock specimen with thickness to diameter ratio of 0.5 is subjected to a load that is spread over the entire thickness of the disc, applying a uniform vertical *line load* diametrically (Figure 3.10). The load is increased to failure, where the sample generally splits along the vertical diametrical plane. The fracture should ideally initiate at the centre and progress vertically towards the loading points. From the theory of elasticity, and assuming the material to be isotropic, the tensile strength of the rock σ_t is given by (Timoshenko, 1934; Hondros, 1959)

$$\sigma_t = \frac{2P}{\pi dt} \tag{3.9}$$

where P = the load at failure, d = specimen diameter and t = specimen thickness. It can be shown that at the centre of the specimen, the minor and the major principal stresses are horizontal and vertical, respectively, at failure. The vertical compressive stress is three times the horizontal tensile stress σ_t given by Equation 3.9.

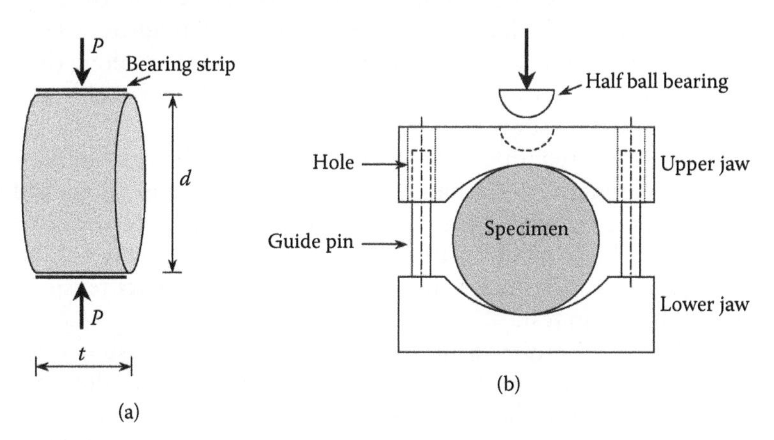

Figure 3.10 Indirect tensile strength test: (a) schematic diagram and (b) loading arrangement.

Mellor and Hawkes (1971) discussed the test procedure in good detail. The standard procedure is discussed in ISRM (1978a) and ASTM D3967. The test works better for brittle materials and has been adopted for concrete, ceramics, cemented soils and asphalt. Note that the recommended *t*/*d* ratio can be different for other materials. In concrete, a length/diameter ratio of 2.0 is recommended for the test specimens. A schematic diagram and the loading arrangement are shown in Figure 3.10a and b, respectively.

3.4.1 Test procedure

The test specimen diameter should be at least of NX core size (54 mm), and the thickness should be approximately equal to half the diameter (ISRM, 1978a). ASTM D3967 allows a *t*/*d* ratio of 0.20–0.75. The loading arrangement suggested by ISRM (1978a) is shown in Figure 3.10b, where the two steel jaws will be in contact with the specimen over an arc length that subtends 10° at the centre, when failure occurs. It is suggested that the radius of the jaws be 1.5 times the specimen radius. The upper jaw has a spherical seating formed by a 25-mm-diameter half ball bearing.

One layer of masking tape is wrapped around the perimeter of the test specimen to cover any irregularities on the contact surface. The specimen is loaded at a constant rate of stress or strain. The measured σ_t is sensitive to the loading rate. The faster the loading, the higher the σ_t (Mellor and Hawkes, 1971). This strain rate effect is commonly seen in soils too. ASTM D3967 suggests that the rate should be selected such that the specimen fails in 1–10 minutes. Considering the scatter, it is often recommended that the test be carried out on 10 specimens and the average value be used.

The state of stress at the *centre* of the specimen is given by a horizontal tensile stress σ_t and a vertical compressive stress that is three times greater in magnitude, both of which are principal stresses (Hondros, 1959). The theoretical basis for Equation 3.9 is that the specimen splits along the vertical diameter. If the fracture plane deviates significantly from being vertical, the test results are questionable.

Indirect tensile strength can be assumed as approximately equal to the direct tensile strength. Goodman (1980) noted that the Brazilian indirect tensile strength test gives a higher value for σ_t than the direct tensile strength test, sometimes by as much as 10 times, especially when there are internal fissures. The fissures in the specimens weaken them in direct tension more than in the Brazilian test.

In the absence of any measurements, σ_t is sometimes assumed to be a small fraction of the uniaxial compressive strength σ_c. A wide range of values from 1/5 to 1/20 have been suggested in the literature, and 1/10 is a good first estimate. The σ_c/σ_t ratios reported by Goodman (1989) along with the σ_c values of several rock types are given in Table 3.7. All σ_t values

Table 3.7 Typical σ_c/σ_t values

Rock type	σ_c *(MPa)*	σ_t *(MPa)*	σ_c/σ_t
Coarse-grained *Nevada granodiorite*	141.1	11.7	12.1
Cedar City tonalite, somewhat weathered quartz monzonite	101.5	6.4	15.9
Fine olivine *Nevada basalt*	148.0	13.1	11.3
Nevada tuff – welded volcanic ash with 19.8% porosity	11.3	1.1	10.0

Source: Goodman (1989).

reported herein are from the Brazilian indirect tensile test. Further descriptions of the rock specimens are given in Goodman (1989).

3.5 POINT LOAD STRENGTH TEST

The origins of the *point load test* can be traced back to the pioneering work of Reichmuth (1968), which was simplified into its present form by Broch and Franklin (1972). It is an index test for strength classification of rocks, where a piece of rock is held between two conical platens of a portable lightweight tester shown in Figure 3.11. Historical development of the point load test and the theoretical background were discussed by Broch and Franklin (1972). The test is rather quick and can be conducted on regular rock cores or irregular rock fragments. The test specimen can be of any of the four forms shown in Figure 3.12. The load is increased to failure and the point load strength index I_s is calculated based on the failure load and the distance D between the cone tips. The *uncorrected point load strength index* I_s is defined as

$$I_s = \frac{P}{D_e^2} \tag{3.10}$$

where D_e is the equivalent diameter of the specimen and I_s is generally given in MPa.

In the diametrical test (Figure 3.12a), $D_e = D$. In axial, block or irregular lump tests (Figures 3.12b, c and d, respectively), the minimum cross-sectional area A of the plane through the platen contact points is computed as $A = WD$. Equating this area to that of a circle, the equivalent diameter D_e is computed as

$$D_e = \sqrt{\frac{4A}{\pi}} = \sqrt{\frac{4WD}{\pi}} \tag{3.11}$$

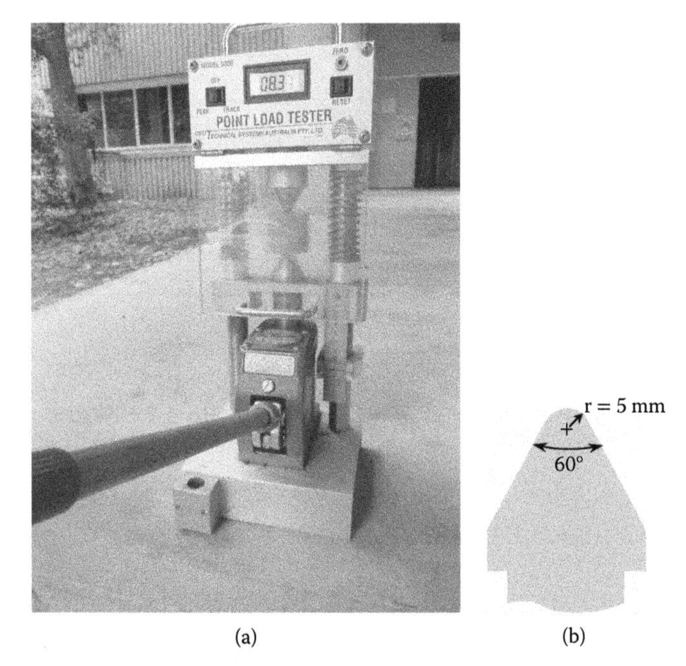

Figure 3.11 (a) Point load tester and (b) conical platen.

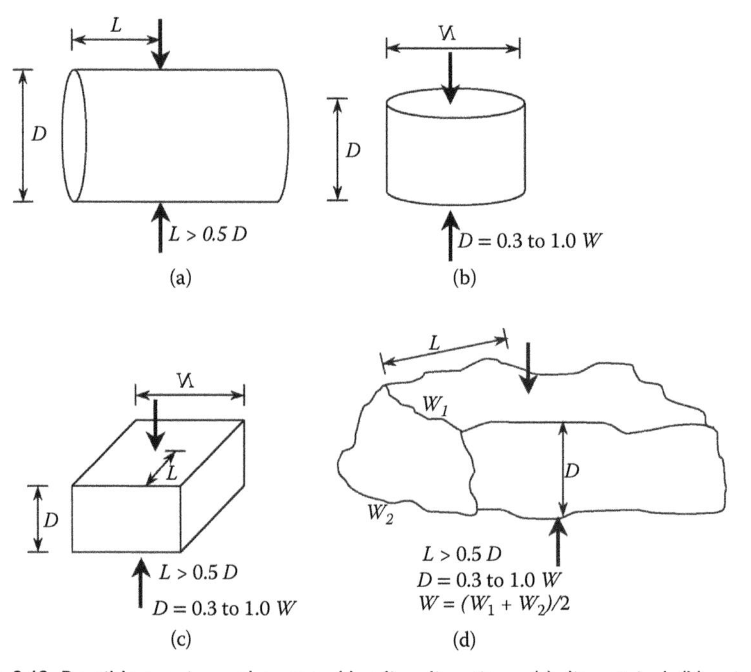

Figure 3.12 Possible specimen shapes and loading directions: (a) diametrical, (b) axial, (c) block and (d) irregular.

It has been observed that I_s increases with D_e, and therefore, it is desirable to have a unique point load index of the rock sample that can be used in rock strength classification. The *size-corrected* point load strength index $I_{s(50)}$ is defined as the value of I_s obtained if D_e is 50 mm. It can be computed as

$$I_{s(50)} = I_s \times \left(\frac{D_e}{50} \right)^{0.45} \tag{3.12}$$

where D_e is in millimetres.

$I_{s(50)}$ is used to classify rocks and is correlated to the strength parameters such as uniaxial compressive strength σ_c or the tensile strength σ_t. A key advantage of a point load test is that it can be carried out on an irregular rock fragment, which is not the case with most other tests where the specimens have to be machined and significant effort is required for preparation. This makes it possible to do the tests at the site on several samples in a relatively short time. Especially during the exploration stage, point load tests are very valuable in making informed decisions and can help in selecting the correct samples for the more sophisticated laboratory tests.

The ratio of uniaxial compressive strength σ_c to $I_{s(50)}$ can be taken as 20–25, but it can vary in the range of 15–50 considering extreme possibilities including anisotropic rocks. Bieniawski (1975) and Broch and Franklin (1972) suggested that $\sigma_c = 24I_{s(50)}$. In spite of the similarities between the point load test and Brazilian indirect tensile strength test, any attempt to derive σ_t from $I_{s(50)}$ should be discouraged (Russell and Wood, 2009). Nevertheless, a crude estimate of Brazilian indirect tensile strength can be obtained as $\sigma_t = 1.25I_{s(50)}$. Point load tests are unreliable for rocks that have uniaxial compressive strength less than 25 MPa (Hoek and Brown, 1997). The test can also be used to quantify the strength anisotropy by the *point load strength anisotropy index* $I_{a(50)}$, defined as the ratio of $I_{s(50)}$ obtained when testing perpendicular and parallel to the planes of weakness. This index is greater than unity when there is anisotropy.

3.5.1 Test procedure

The standard test procedure is described in ISRM (1985) and ASTM D5731. The test is carried out on a specimen that can be of any of the four forms shown in Figure 3.12, with *equivalent diameter* D_e of 30–85 mm. It is held between the two conical ends of the point load tester, and the load is applied to failure. The loading is rather quick so that the specimen fails in 10–60 seconds. It is recommended that the test be carried out on at least 10 specimens (more if anisotropic or heterogeneous) where the highest two and the lowest two values are discarded and the average value of the remaining specimens be reported as the point load index. Any specific test where the failure does not extend to the full depth should be rejected. A typical point load test datasheet is shown in Table 3.8.

Table 3.8 Point load test data

No.	*Type*	W *(mm)*	*D (mm)*	*P (kN)*	D_e *(mm)*	I_s *(MPa)*	$I_{s(50)}$ *(MPa)*
1	i ⊥	30.4	17.2	2.687	25.8	4.04	~~3.00~~
2	i ⊥	16.0	8.0	0.977	12.8	5.99	3.24
3	i ⊥	19.7	15.6	1.962	19.8	5.01	3.30
4	i ⊥	35.8	18.1	3.641	28.7	4.41	3.44
5	i ⊥	42.5	29.0	6.119	39.6	3.90	3.51
6	i ⊥	42.0	35.0	7.391	43.3	3.95	~~3.70~~
7	b ⊥	44	21	4.600	34.3	3.91	3.30
8	b ⊥	40	30	5.940	39.1	3.89	3.48
9	b ⊥	19.5	15	2.040	19.3	5.48	~~3.57~~
10	b ⊥	33	16	2.870	25.9	4.27	~~3.18~~
11	d//	–	49.93	5.107	49.93	2.05	2.05
12	d//	–	49.88	4.615	49.88	1.85	1.85
13	d//	–	49.82	5.682	49.82	2.29	~~2.29~~
14	d//	–	49.82	4.139	49.82	1.67	~~1.66~~
15	d//	–	49.86	4.546	49.86	1.83	1.83
16	d//	–	25.23	1.837	25.23	2.89	2.12
17	d//	–	25.00	1.891	25	3.03	2.21
18	d//	–	25.07	2.118	25.07	3.37	~~2.47~~
19	d//	–	25.06	1.454	25.06	2.32	~~1.70~~
20	d//	–	25.04	1.540	25.04	2.46	1.80

Mean $I_{s(50)\perp}$	3.38 MPa
Mean $I_{s(50)//}$	1.98 MPa
$I_{a(50)}$	1.71

a = axial

b = block

d = diametrical

i = irregular lump

⊥ = loaded perpendicular to plane of weakness

// = loaded parallel to plane of weakness

Source: Adapted from ISRM, *Int. J. Rock Mech. Min. Sci. & Geomech. Abstr.*, 22, 51–60, 1985.

3.6 SLAKE DURABILITY TEST

Rocks are generally weaker when wet than dry due to the presence of water in the cracks and its subsequent reaction to the applied loads during the tests. Repeated wetting and drying, which happens often in service, can weaken the rock significantly. *Slaking* is a process of disintegration of an aggregate when in contact with water. The *slake durability index* quantifies the resistance of a rock to wetting and drying cycles and is seen as a measure of the *durability* of the rock. This is mainly used for weak rocks such as shales, mudstones, claystones and siltstones. The slake durability test is an index test that was first proposed by Franklin and Chandra (1972) doing their PhD and MSc work, respectively, at London University in 1970.

Figure 3.13 shows the slake durability apparatus, which consists of two rotating sieve mesh drums immersed in a water bath. Ten rock lumps, each weighing 40–60 g, are placed in the drum and rotated for 10 minutes, allowing for disintegrated fragments to leave the drum through the 2-mm sieve mesh. The remaining fragments in the drum are dried and weighed. Gamble (1971), a PhD student at University of Illinois, United States, suggested this be repeated over a second cycle of slaking. The dry mass of the sample remaining in the drum at the end of second cycle, expressed as a percentage of the *original dry mass* in the drum at the beginning of the test, is known as the *second-cycle slake durability index* I_{d2}, which varies in the range of 0–100%. For samples that are highly susceptible to slaking, I_{d2} is close to zero and for very durable rocks, it is close to 100%.

The *first-cycle slake durability index* I_{d1} is defined as

$$I_{d1} = \frac{m_2}{m_1} \times 100 \tag{3.13}$$

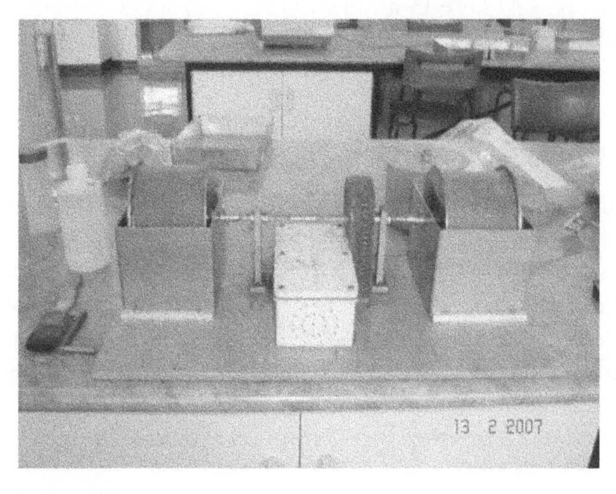

Figure 3.13 Slake durability apparatus.

Table 3.9 Durability classification based on slake durability index

Durability	I_{d1}	I_{d2}
Very high	>99	98–100
High	98–99	95–98
Medium high	95–98	85–95
Medium	85–95	60–85
Low	60–85	30–60
Very low	<60	0–30

Source: Gamble, J.C., *Durability—Plasticity classification of shales and other argillaceous rocks*, PhD thesis, University of Illinois at Urbana-Champaign, IL, 1971.

The *second-cycle slake durability index* I_{d2} is defined as

$$I_{d2} = \frac{m_3}{m_1} \times 100 \tag{3.14}$$

Here, m_1 = dry mass of the original lumps in the drum, m_2 = dry mass of the material retained in the drum after the first cycle and m_3 = dry mass of the material retained after the second cycle.

The second-cycle slake durability index I_{d2} is the one that is commonly used as a measure of rock durability. Only in rocks that are classified as very low in durability with $I_{d2} < 10\%$, it is recommended to include I_{d1} as well. A durability classification of rocks, based on slake durability index as proposed by Gamble (1971), is given in Table 3.9. This is slightly different to what is proposed by Franklin and Chandra (1972), who did not distinguish between the two cycles and used a single slake durability index I_d based on the first cycle. ASTM D4644 and ISRM (1979b) suggest reporting I_{d2} as the slake durability index. For rocks of higher durability, three or more cycles (i.e., I_{d3}, I_{d4} etc.) may be appropriate.

3.6.1 Test procedure

The standard procedure for the slake durability test is described in ISRM (1979b) and ASTM D4644. A representative sample of 10 rock lumps, each with a mass of 40–60 g, giving a total mass of 450–550 g, is dried and placed within the drum. The corners of the lumps should be rounded off so that they are approximately spherical. The drum is partly submerged in the slaking fluid (see Figure 3.13), which can be tap water, seawater and so on, to simulate the service environment. For each cycle, the drum is rotated at a standard rate of 20 rev/min for 10 minutes. Generally, only I_{d2} is reported to the nearest 0.1%. Only when I_{d2} is less than 10%, it is suggested to report I_{d1} as well. A typical slake durability test datasheet is shown in Table 3.10.

Table 3.10 Slake durability test datasheet

Sample no.	*Porcellanite2*	*Porcellanite7*	*Claystone1*	*Claystone3*	*Claystone8*
Mass of drum + dry sample (m_1), g	1476	1457	1464	1493	1503
Mass of drum + dry sample after first cycle (m_2), g	1472	1452	1125	1114	1103
Mass of drum + dry sample after second cycle (m_3), g	1467	1446	1013	1004	1009
Mass of drum (m_4), g	971	970	968	969	968
Second-cycle slake durability index, I_{d2}	98.2	97.7	9.1	6.7	7.7
First-cycle slake durability index, I_{d1}	99.2	99.0	31.7	27.7	25.2
Mass of drum + dry sample after third cycle, g (only if required)	1464	1443	–	–	–
Duration of third cycle (if not 10 minutes)					
Third-cycle slake durability index, I_{d3}	97.6	97.1	–	–	–
Mass of drum + dry sample after fourth cycle, g (only if required)	1468.0	1447.0			
Duration of fourth cycle (if not 10 minutes)	30 minutes	30 minutes			
Fourth-cycle slake durability index, I_{d4}	98.4	97.9			
Slaking fluid	Seawater			Tap water	
Temperature of slaking fluid	26°C	26°C	27°C	27°C	27°C

The usefulness of the slake durability test is limited to relatively weak rocks such as shales, mudstones and other highly weathered rocks.

3.7 SCHMIDT HAMMER TEST

The Schmidt (1951) hammer (Figure 3.14) was originally developed in 1948 for testing the hardness of concrete. It is a simple, portable and inexpensive device that gives the rebound hardness value R for an intact rock specimen in

Figure 3.14 Schmidt hammer test.

the laboratory or the rock mass in situ. The test is generally non-destructive for rocks of at least moderate strength, and therefore, the same specimen can be used for other tests. ASTM D5873 and ISRM (1978b) recommend this test for rocks with UCS of 1–100 MPa and 20–150 MPa, respectively. This is a popular index test on rocks, and the rebound hardness R has been correlated with rock properties such as UCS and E. The ISRM suggested method was revised by Aydin (2009).

The hammer consists of a spring-loaded metal piston that is released when the plunger is pressed against the rock surface. The impact of the piston onto the plunger transfers the energy to the rock. How much of this energy is recovered depends on the hardness of the rock and is measured by the rebound height of the piston. The harder the surface, the shorter the penetration time (i.e., smaller impulse and less energy loss) and hence the greater the rebound. Rebound hardness R is a number that varies in the range of 0–100.

Two types of Schmidt hammers are commonly used. They are L-type with an impact energy of 0.735 N·m and N-type with an impact energy of 2.207 N·m. The measured rebound hardness is denoted by R_L and R_N respectively. Other few notations used in the literature for rebound hardness are H_R, N, SRH and so on. Prior to 2009, ISRM recommended only L-type hammers; now they are both allowed (Aydin, 2009). N-type hammers were mostly used for concrete. However, they are less sensitive to surface irregularities and suit field applications. ASTM does not specify the type of hammer.

3.7.1 Test procedure

A Schmidt hammer must be calibrated first, using a calibration test anvil supplied by the manufacturer based on the average of 10 readings. A correction factor is computed as

$$\mathrm{CF} = \frac{\text{Specified standard value of the anvil}}{\text{Average of the 10 readings on the anvil}} \tag{3.15}$$

and it has to be applied to all future readings. This factor is to account for the spring losing its stiffness with time. For L-type hammers, the test specimen must be of at least NX (54 mm) core size, with length greater than 100 mm (ISRM). ASTM suggests a minimum length of 150 mm. For N-type hammers, ISRM suggests 84 mm diameter or larger cores (Aydin, 2009). The hammer should be used vertically upwards, horizontally or vertically downwards with $\pm 5°$ tolerance. ISRM recommends 20 readings at different locations with an option to stop when the subsequent 10 readings differ by less than 4. ASTM recommends 10 readings. ISRM (1978b) suggests using the average of the top 10 readings. ASTM recommends discarding the readings that differ from the average by more than 7 and averaging the rest. The revised ISRM (Aydin, 2009) suggests not discarding any data and presenting the values as a histogram with mean, median, mode and range.

3.8 TRIAXIAL TEST

As a first approximation, it can be assumed that rocks, like most geomaterials, follow the *Mohr–Coulomb* failure criterion given by

$$\tau_f = c + \sigma \tan\phi \tag{3.16}$$

where τ_f = shear strength (or shear stress at failure on the failure plane), σ = normal stress on the failure plane, c = cohesion and ϕ = friction angle. Cohesion and friction angle are the shear strength parameters of the rock and are constants. Thus, it can be seen from Equation 3.16 that τ_f is proportional to σ. In terms of major and minor principal stresses at failure, Equation 3.16 can be written as

$$\sigma_1 = \left(\frac{1+\sin\phi}{1-\sin\phi}\right)\sigma_3 + 2c\left(\frac{1+\sin\phi}{1-\sin\phi}\right)^{0.5} \tag{3.17}$$

There are also other *failure criteria* for rocks such as Hoek-Brown, where the failure envelope is nonlinear.

Similar to the triaxial tests on soils, here too cylindrical rock specimens are subjected to different confining pressures and loaded axially to failure (Figure 3.15a and b). The only difference is that the loads and pressures are much higher. The test procedure suggested by ISRM (1983) does not have a provision for pore water pressure or drainage measurements. It is similar to an unconsolidated undrained triaxial test on a clay specimen. Only the procedure for an individual test is described here. The procedures for a multiple failure state test, similar to staged test and a continuous failure state test, are given in ISRM (1983).

3.8.1 Test procedure

The test specimen diameter should be at least of NX core size (54 mm), and the length should be approximately equal to two to three times the diameter. The test specimens should be cut and prepared using clean water.

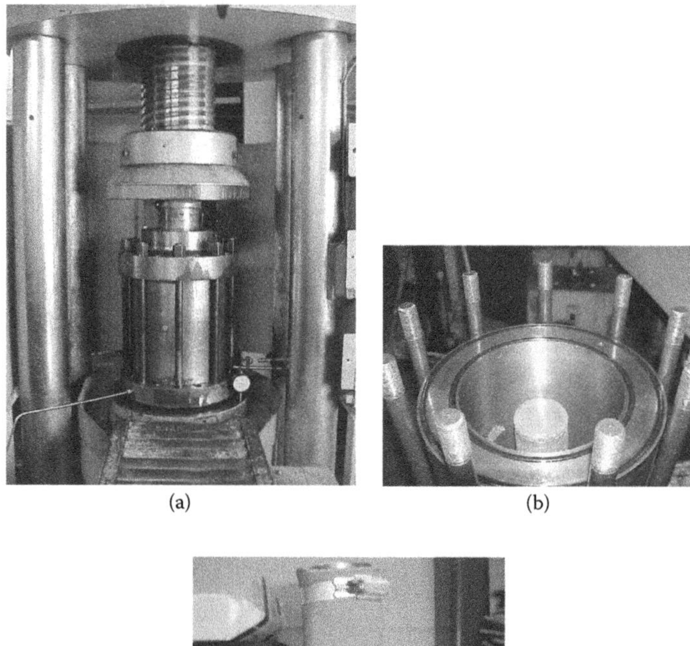

(a) (b)

(c)

Figure 3.15 Rock triaxial test: (a) triaxial test in progress, (b) triaxial cell interior with specimen and (c) rock specimen enclosed in membrane.

The ends of the test specimens shall be flat to ±0.01 mm and be parallel to each other and at right angles to the longitudinal axis. The sides of the specimens shall be smooth and free of abrupt irregularities and straight, within 0.3 mm over the full length of the specimen. The diameter of the specimen should be at least 10 times larger than the largest mineral grain present. Use of capping material or end surface treatment is not permitted.

The specimen is enclosed in a flexible impervious membrane (Figure 3.15c) to prevent the confining fluid from entering the specimen pores. Sometimes, it is required to make customised membranes that suit the different core diameters. Oil is generally used as the confining fluid and the confining pressure (σ_3) is increased to desired levels. The vertical stress ($\Delta\sigma$) on the specimen is increased at a constant stress/strain rate (e.g., 0.5–1.0 MPa/s) until failure occurs, ideally in 5–15 minutes. The vertical stress at failure (σ_1) is given by $\sigma_3 + \Delta\sigma$.

3.9 EMPIRICAL CORRELATIONS

There are several empirical correlations interrelating the intact rock parameters such as uniaxial compressive strength σ_c, indirect tensile strength σ_t, point load strength index $I_{s(50)}$ and so on. Some of the correlations between the uniaxial compressive strength and the indirect tensile strength are summarised in Table 3.11. Correlations between the uniaxial compressive strength and the point load strength index are summarised in Table 3.12.

Gunsallus and Kulhawy (1984) carried out these tests on rock specimens of dolostones (predominantly dolomite), sandstones and limestones in the United States, representing eight different rock types and assessed the different correlations reported in literature to find that the two popular correlations $\sigma_c = 10\sigma_t$ and $\sigma_c = 24I_{s(50)}$ work quite well.

Table 3.11 σ_c–σ_t Correlations

Correlation	*Reference*	*Comments*
$\sigma_c = 10.5\sigma_t + 1.2$	Hassani et al. (1979)	
$\sigma_c = 3.6\sigma_t + 15.2$	Szlavin (1974)	United Kingdom; 229 tests
$\sigma_c = 2.84\sigma_t - 3.34$	Hobbs (1964)	Mudstone, sandstone and limestone
$\sigma_c = 12.4\sigma_t - 9.0$	Gunsallus and Kulhawy (1984)	Dolostone, sandstone and limestone from the United States
$\sigma_c = 10\sigma_t$	Broch and Franklin (1972)	

Table 3.12 σ_c–$I_{s(50)}$ Correlations

Correlation	*Reference*	*Comments*
$\sigma_c = 24I_{s(50)}$	Broch and Franklin (1972)	
$\sigma_c = 24I_{s(50)}$	Bieniawski (1975)	Sandstone, South Africa
$\sigma_c = 29I_{s(50)}$	Hassani et al. (1980)	Sedimentary rocks, United Kingdom
$\sigma_c = 14.5I_{s(50)}$	Forster (1983)	Dolerite and sandstone
$\sigma_c = 12.5I_{s(50)}$	Chau and Wong (1996)	Hong Kong rocks
$\sigma_c = 16I_{s(50)}$	Read et al. (1980)	Basalt
$\sigma_c = 20I_{s(50)}$	Read et al. (1980)	Sedimentary rocks, Australia
$\sigma_c = 23.4I_{s(50)}$	Singh and Singh (1993)	Quartzite, India
$\sigma_c = 15.3I_{s(50)} + 16.3$	D'Andrea et al. (1964)	Range of rock types
$\sigma_c = 16.5I_{s(50)} + 51.0$	Gunsallus and Kulhawy (1984)	Dolostone, sandstone and limestone from the United States
$\sigma_c = 9.3I_{s(50)} + 20.04$	Grasso et al. (1992)	
$\sigma_c = 23I_{s(54)} + 13$	Cargill and Shakoor (1990)	Mostly from United States and some from Canada; 54-mm-diameter cores

3.10 SUMMARY

1. A UCS of 1 MPa is the cut-off between soils and rocks.
2. Laboratory tests are generally carried out on intact rock specimens, which will not reflect the discontinuities present within the rock mass.
3. UCS is the most used strength parameter in the designs and analysis of rocks.
4. It is difficult to carry out a proper tensile strength test on rocks. The Brazilian indirect tensile test is a simple way around this problem. How close the estimated tensile strength is to the real value is the million dollar question.
5. In the Brazilian indirect tensile test, tensile failure is induced in the rock specimen by applying a vertical compressive load diametrically.
6. The advantage of point load test is that it can be tested on irregular-shaped specimens and gives a quick estimate of the point load strength index. The simple apparatus can be taken to the site where several specimens can be tested within few minutes, which will be of good value in preliminary assessments.
7. The Schmidt hammer test is not recommended for very weak or very hard rocks. It is a non-destructive test that can be carried out on rock cores in the laboratory or in the field outcrops. It gives a dimensionless empirical relative hardness number in the range of 0–100.
8. Triaxial tests are effective in assessing the strength variation with confining pressures.

Review Exercises

1. State whether the following are true or false.
   i. Uniaxial compressive strength is the same as UCS.
   ii. The point load strength index is a dimensionless number.
   iii. The larger the slake durability index, the higher the durability of the rock in wetting and drying.
   iv. In the slake durability test, I_{d2} is always less than I_{d1}.
   v. In a UCS test, the larger the specimen diameter, the larger the strength.
   vi. In a UCS test, the faster the rate of loading, the lower the strength.
   vii. The larger the core size, the larger the uniaxial compressive strength.
2. Circle the correct answer.
   i. Which of the following rock cores is larger in diameter?
      a. AQ
      b. BQ
      c. HQ
      d. NQ
   ii. Which of the following rock core diameters is the minimum recommended size for most laboratory tests?
      a. AX
      b. BX
      c. EX
      d. NX
   iii. The typical range for the uniaxial compressive strength of rocks is
      a. 1–400 kPa
      b. 1–400 MPa
      c. 1–400 GPa
      d. None of these
   iv. Which of the following can be a typical value for the *E*/UCS ratio of a rock?
      a. 3
      b. 30
      c. 300
      d. 3000
   v. Which of the following tests require the most sample preparation?
      a. Slake durability test
      b. Point load test
      c. UCS test
      d. Schmidt hammer test
   vi. Which of the following tests require the least sample preparation?
      a. UCS test
      b. Indirect tensile strength test
      c. Point load test
      d. Direct tensile strength test

vii. Which of the following is the preferred aspect (length/diameter) ratio for a Brazilian indirect tensile strength test specimen?
   a. 3
   b. 2
   c. 1
   d. 0.5
viii. Which one of the following tensile strengths does the Brazilian indirect tensile strength test measure?
   a. At the centre
   b. At the top of the diameter
   c. At the bottom of the diameter
   d. Average value for the entire specimen volume
ix. Which of the following parameters (in MPa) would be the smallest?
   a. σ_c from a UCS test
   b. σ_t from a Brazilian indirect tensile strength test
   c. $I_{s(50)}$ from a point load strength test

3. In a 1500-mm rock core run, the following rock pieces were recovered from a borehole: 53 mm, 108 mm, 125 mm, 75 mm, 148 mm, 320 mm, 68 mm, 145 mm, 35 mm and 134 mm. Find the RQD and the core recovery ratio.
   Answer: 65%, 81%
4. For a cylindrical rock specimen subjected to an axial load (e.g., UCS), neglecting higher order terms of strains, show that the volumetric strain ε_{vol} is given by

$$\varepsilon_{vol} = \varepsilon_a + 2\varepsilon_d$$

   where ε_a = axial strain and ε_d = diametral strain.
5. Point load tests were carried out on two sedimentary rock specimens of 54 mm diameter (NX core), as shown the following figure The loads $P_\perp$ and $P_{//}$ at failure are 6.28 kN and 4.71 kN, respectively. Find the point load strength index $I_{s(50)}$ in the two directions and compute the point load strength anisotropy index $I_{a(50)}$.

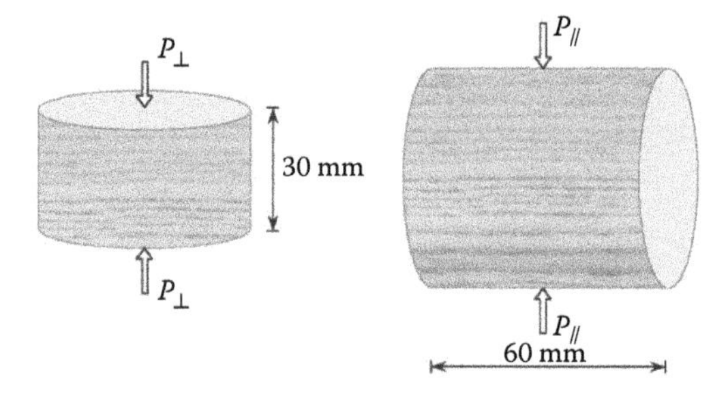

   Answer: 2.92 MPa, 1.68 MPa; 1.74

6. Surf the web and do a research on the following and explain them in less than 100 words each.
   a. Wire line drilling
   b. Triple-tube sampling
   c. Types of drilling in rocks

REFERENCES

ASTM D3967-08. Standard test method for splitting tensile strength of intact rock core specimens.

ASTM D4543-08. Standard practices for preparing rock core as cylindrical test specimens and verifying conformance to dimensional and shape tolerances.

ASTM D4644-08. Standard test method for slake durability of shales and similar weak rocks.

ASTM D5731-08. Standard test method for determination of the point load strength index and application to rock strength classifications.

ASTM D5873-05. Standard test method for determination of rock hardness by rebound hammer method.

ASTM D6032-08. Standard test method for determining rock quality designation (RQD) of rock core.

ASTM D7012-07e1. Standard test method for compressive strength and elastic moduli of intact rock core specimens under varying states of stress and temperatures.

Aydin, A. (2009). ISRM suggested method for determination of the Schmidt hammer rebound hardness: Revised version. *International Journal of Rock Mechanics and Mining Sciences*, Vol. 46, No. 3, pp. 627–634.

Bieniawski, Z.T. (1975). The point load test in geotechnical practice. *Engineering Geology*, Vol. 9, No. 1, pp. 1–11.

Broch, E. and Franklin, J.A. (1972). The point load strength test. *International Journal of Rock Mechanics and Mining Sciences*, Vol. 9, No. 6, pp. 669–697.

Brown, E.T. (Ed.) (1981). *Rock Chracterisation Testing and Monitoring – ISRM Suggested Methods*. Pergamon Press, Oxford, 211pp.

Cargill, J.S. and Shakoor, A. (1990). Evaluation of empirical methods for measuring the uniaxial strength of rock. *International Journal of Rock Mechanics and Mining Sciences & Geomechanics Abstracts*, Vol. 27, No. 6, pp. 495–503.

Chau, K.T. and Wong, R.H.C. (1996). Uniaxial compressive strength and point load strength. *International Journal of Rock Mechanics and Mining Sciences & Geomechanics Abstracts*, Vol. 33, No. 2, pp. 183–188.

D'Andrea, D.V., Fisher, R.L. and Fogelson, D.E. (1964). Prediction of compression strength from other rock properties. *Colorado School of Mines Quarterly*, Vol. 59, No. (4B), pp. 623–640.

Deere, D.U. (1964). Technical description of rock cores for engineering purposes. *Rock Mechanics and Engineering Geology*, Vol. 1, pp. 17–22.

Deere, D.U. and Miller, R.P. (1966). Engineering classification and index properties of intact rock. *Report AFWL-TR-65-116, Air Force Weapon Laboratory (WLDC)*, Kirtland Airforce Base, New Mexico, 308pp.

Forster, I.R. (1983). The influence of core sample geometry on the axial point-load test. *International Journal of Rock Mechanics and Mining Science & Geomechanics Abstracts*, Vol. 20, No. 6, pp. 291–295.

Franklin, J.A. and Chandra, J.A. (1972). The slake durability test. *International Journal of Rock Mechanics and Mining Sciences*, Vol. 9, No. 3, pp. 325–341.

Gamble, J.C. (1971). *Durability – Plasticity classification of shales and other argillaceous rocks*, PhD thesis, University of Illinois at Urbana-Champaign, IL, 159pp.

Gercek, H. (2007). Poisson's ratio values for rocks. *International Journal of Rock Mechanics and Mining Sciences*, Vol. 44, No. 1, pp. 1–13.

Goodman, R.E. (1989). *Introduction to Rock Mechanics*. 2nd edition, Wiley, New York.

Grasso, P., Xu, S. and Mahtab, A. (1992). Problems and promises of index testing of rocks. *Proceedings of the 33rd US Symposium of Rock Mechanics*, Santa Fe, New Mexico, Balkema, Rotterdam, pp. 879–888.

Gunsallus, K.L. and Kulhawy, F.H. (1984). A comparative evaluation of rock strength measures. *International Journal of Rock Mechanics and Mining Sciences & Geomechanics Abstracts*, Vol. 21, No. 5, pp. 233–248.

Handin, J. (1966). Strength and ductility. *Handbook of Physical Contacts*, Ed. S.P. Clark, Geological Society of America, New York, pp. 223–289.

Hassani, F.P., Scoble, M.J. and Whittaker, B.N. (1980). Application of point load index test to strength determination of rock and proposals for new size-correction chart. *Proceedings of 21st US Symposium on Rock Mechanics*, Ed. D.A. Summers, University of Missouri Press, Rolla, MO, pp. 543–564.

Hassani, F.P., Whittaker, B.N. and Scoble, M.J. (1979). Strength characteristics of rocks associated with opencast coal mining in UK. *Proceedings of 20th U.S. Symposium on Rock Mechanics*, Austin, TX, pp. 347–356.

Hawkes, I. and Mellor, M. (1970). Uniaxial testing in rock mechanics laboratories. *Engineering Geology*, Vol. 4, No. 3, pp. 179–285.

Hobbs, D.W. (1964). Simple method for assessing uniaxial compressive strength of rock. *International Journal of Rock Mechanics and Mining Sciences & Geomechanics Abstracts*, Vol. 1, No. 1, pp. 5–15.

Hoek, E. and Brown, E.T. (1980). *Underground Excavations in Rock*. Institution of Mining and Metallurgy, London, 527pp.

Hoek, E. and Brown, E.T. (1997). Practical estimates of rock mass strength. *International Journal of Rock Mechanics and Mining Sciences & Geomechanics Abstracts*, Vol. 34, No. 8, pp. 1165–1186.

Hoek, E. and Diederichs, M.S. (2006). Empirical estimation of rock mass modulus. *International Journal of Rock Mechanics and Mining Sciences*, Elsevier, Vol. 43, No. 2, pp. 203–215.

Hondros, G. (1959). The evaluation of Poisson's ratio and the modulus of materials of a low tensile resistance by the Brazilian (indirect tensile) test with particular reference to concrete. *Australian Journal of Applied Science*, Vol. 10, No. 3, pp. 243–268.

Hudson, J.A. (1989). *Rock Mechanics Principles in Engineering Practice*. Butterworths, London.

ISRM (1978a). International Society of Rock Mechanics, Commission on Standardisation of Laboratory and Field Tests. Suggested methods for determining tensile strength of rock materials. *International Journal of Rock Mechanics and Mining Sciences & Geomechanics Abstract*, Vol. 15, No. 3, pp. 99–103.

ISRM (1978b). International Society of Rock Mechanics, Commission on Standardisation of Laboratory and Field Tests. Suggested methods for determining hardness and abrasiveness of rocks. *International Journal of Rock Mechanics and Mining Sciences & Geomechanics Abstract*, Vol. 15, No. 3, pp. 89–97.

ISRM (1978c). International Society of Rock Mechanics, Commission on Standardisation of Laboratory and Field Tests. Suggested methods for the quantitative description of rock discontinuities in rock masses. *International Journal of Rock Mechanics and Mining Sciences & Geomechanics Abstracts*, Vol. 15, No. 6, pp. 319–368.

ISRM (1979a). International Society of Rock Mechanics, Commission on Standardisation of Laboratory and Field Tests. Suggested methods for determination of the uniaxial compressive strength of rock materials. *International Journal of Rock Mechanics and Mining Sciences & Geomechanics Abstract*, Vol. 16, No. 2, pp. 135–140.

ISRM (1979b). International Society of Rock Mechanics, Commission on Standardisation of Laboratory and Field Tests. Suggested methods for determining water content, porosity, density, absorption and related properties and swelling and slake durability index properties. *International Journal of Rock Mechanics and Mining Sciences & Geomechanics Abstracts*, Vol. 16, No. 2, pp. 143–156.

ISRM (1983). International Society of Rock Mechanics, Commission on Testing Methods. Suggested method for determining the strength of rock materials in triaxial compression. *International Journal of Rock Mechanics and Mining Sciences & Geomechanics Abstract*, Vol. 20, No. 6, pp. 285–290.

ISRM (1985). International Society of Rock Mechanics, Commission on Testing Methods. Suggested method for determining point load strength. *International Journal of Rock Mechanics and Mining Sciences & Geomechanics Abstract*, Vol. 22, No. 2, pp. 51–60.

Mellor, M. and Hawkes, I. (1971). Measurement of tensile strength by diametral compression of discs and annuli. *Engineering Geology*, Vol. 5, No. 3, pp. 173–225.

Russell, A.R. and Wood, D.M. (2009). Point load tests and strength measurements for brittle spheres. *International Journal of Rock Mechanics and Mining Sciences*, Vol. 46, No. 2, pp. 272–280.

Peck, R.B., Hanson, W.E. and Thornburn, T.H. (1974). *Foundation Engineering*. 2nd edition, John Wiley & Sons, New York.

Read, J.R.L., Thornten, P.N. and Regan, W.M. (1980). A rational approach to the point load test. *Proceedings of ANZ Geomechanics Conference*, Vol. 2, pp. 35–39.

Reichmuth, D.R. (1968). Point load testing of brittle materials to determine tensile strength and relative brittleness, Status of Practical Rock Mechanics. *Proceedings of 9th Symposium of Rock Mechanics, Colorado School of Mines*, Boulder, CO, pp. 134–159.

Schmidt, E. (1951). A non-destructive concrete tester. *Concrete*, Vol. 59, No. 8, pp. 34–35.
Singh, V.K. and Singh, D.P. (1993). Correlation between point load index and compressive strength for quartzite rocks. *Geotechnical and Geological Engineering*, Vol. 11, pp. 269–272.
Szlavin, J. (1974). Relationship between some physical properties of rock determined by laboratory tests. *International Journal of Geomechanics and Mining Sciences & Geomechanics Abstracts*, Vol. 11, No. 2, pp. 57–66.
Timoshenko, S. (1934). *Theory of Elasticity*. McGraw Hill, New York.

Chapter 4

Rock mass classification

4.1 INTRODUCTION

When a site investigation is carried out, cylindrical rock cores are collected from boreholes for their identification, laboratory tests and classification. What we learn from these rock cores is only part of the story; the situation can be very different in the larger rock mass in situ, thanks to the discontinuities present in the rock mass in the form of joints, faults and bedding planes. These are the planes of weakness, which become the weakest links and can cause instability.

The rock cores are intact rock specimens that are so small that they are often free of discontinuities. Even when they break along discontinuities, we trim them further to have a 'joint-free' core for the laboratory tests. On the other hand, the larger rock mass may have one or more sets of discontinuities that can have a significant influence on stability, which is not reflected on the intact rock specimen. The strength of the intact rock is only one of the key parameters used in classifying the rock mass. The different laboratory tests discussed in Chapter 3 are for intact rock specimens, with no reflection on the extent of discontinuities present within the rock mass. The laboratory test data are only used here to get the big picture relating to the much larger rock mass. It is important to understand the difference between the rock mass and the intact rock.

As we discussed when looking at kinematic analysis in Chapter 2, the orientations of the discontinuities can play a significant role in the stability of rock slopes and underground openings. In the same rock mass, the orientation of the discontinuity sets can be favourable or unfavourable, depending on how the facility (e.g. tunnel) is located with respect to the orientations of the discontinuities. There are several variables (e.g. orientation, spacing etc.) associated with the discontinuities in a rock mass, all of which are relevant in rock mass classification. This chapter discusses the different ways of classifying the rock mass with due consideration to the above variables, including the intact rock strength.

4.2 INTACT ROCK AND ROCK MASS

Figure 4.1a shows a schematic diagram of a rock mass with two sets of discontinuities, and an intact rock specimen that is typically tested in the laboratory. The stability of the rock mass under a specific loading condition (e.g. foundation or tunnelling) can be very different from the stability of the intact rock specimen, thanks to the discontinuities. Due to the presence of discontinuities, the rock mass is weaker than the intact rock specimen, showing lower strength and stiffness (see Figure 4.1b). In addition, the rock mass is more permeable, with the discontinuities allowing greater access to water, which can make the rock mass even weaker. Water reduces the friction along the discontinuities, and the increased pore water pressure reduces the effective stresses and hence the shear strength.

The stability of the rock mass is thus governed by the properties of the intact rock as well as the relative ease at which the rock pieces (or blocks) can slide, rotate or topple. This in turn is influenced by the dimensions of the individual blocks and the frictional characteristics at the joints that separate the blocks. We will see in the following sections that the rock mass is generally characterised based on the properties of the intact rock, block size and the frictional characteristics of the joint. The frictional characteristics include the roughness profile of the joint surface and the quality of the infill material.

Discontinuity is a generic term used to describe a fault, joint, bedding plane, foliation, cleavage or schistosity. Fault is a planar fracture along which noticeable movement has taken place. Joints are filled or unfilled fractures within the rock mass that do not show any sign of relative movement (Figure 4.2). Bedding planes are formed when the sediments are deposited

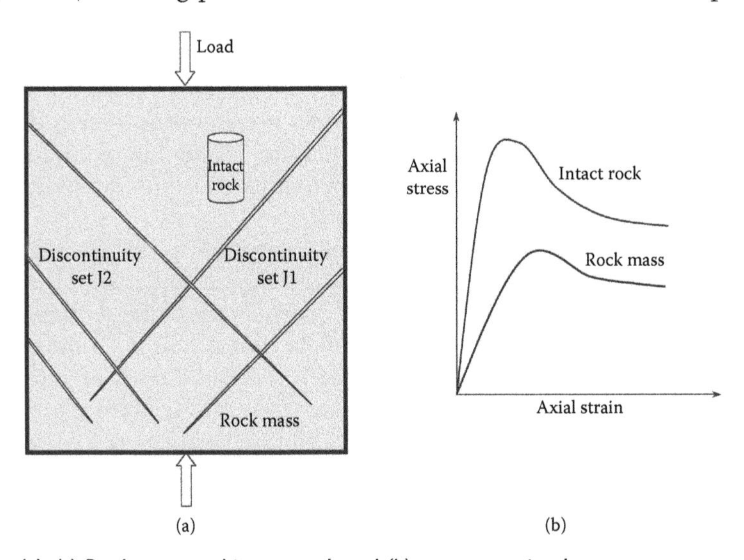

Figure 4.1 (a) Rock mass and intact rock and (b) stress–strain plot.

(a)

(b)

Figure 4.2 Joints: (a) filled and (b) unfilled.

in the rock formation process, creating planes of weakness, which are not necessarily horizontal. They are common in sedimentary rocks. Foliation occurs in metamorphic rocks where the rock-forming minerals exhibit platy structure or banding, thus developing planes of weakness. Cleavages are planes of weaknesses that occur often as parallel layers and are formed in a metamorphic process. Schistosity is a type of cleavage seen in metamorphic rocks such as schists and phyllites, where the rocks tend to split along parallel planes of weakness.

Table 4.1 shows the classification of soils and rocks on the basis of the uniaxial compressive strength, as recommended by the International Society for Rock Mechanics (ISRM) (1978). Also shown in the table are

Table 4.1 Classification of soil and rock strengths

Grade	*Description*	*Field identification*	σ_c *or* q_u *(MPa)*	*Rock types*
S1	Very soft clay	Easily penetrated several inches by fist.	<0.025	
S2	Soft clay	Easily penetrated several inches by thumb.	0.025–0.05	
S3	Firm clay	Can be penetrated several inches by thumb with moderate effort.	0.05–0.10	
S4	Stiff clay	Readily indented by thumb, but penetrated only with great effort.	0.1–0.25[a]	
S5	Very stiff clay	Readily indented by thumbnail.	0.25[a]–0.50[a]	
S6	Hard clay	Indented with difficulty by thumbnail.	>0.5[a]	
R0	Extremely weak rock	Indented by thumbnail.	0.25–1.0	Stiff fault gouge
R1	Weak rock	Crumbles under firm blows with point of geological hammer; can be peeled by pocketknife.	1–5	Highly weathered or altered rock
R2	Weak rock	Can be peeled by a pocketknife with difficulty; shallow indentations made by firm blow with a point of geological hammer.	5–25	Chalk, rock salt, potash
R3	Medium strong rock	Cannot be scraped or peeled with a pocketknife; specimen can be fractured with a single firm blow of a geological hammer.	25–50	Claystone, coal, concrete, schist, shale, siltstone

Table 4.1 Classification of soil and rock strengths (*Continued*)

Grade	*Description*	*Field identification*	σ_c *or* q_u *(MPa)*	*Rock types*
R4	Strong rock	Specimen requires more than one blow by geological hammer to fracture it.	50–100	Limestone, marble, phyllite, sandstone, schist, shale
R5	Very strong rock	Specimen requires many blows of geological hammer to fracture it.	100–250	Amphibiolite, sandstone, basalt, gabbro, gneiss, granodiorite, limestone, marble, rhyolite, tuff
R6	Extremely strong rock	Specimen can only be chipped by a geological hammer.	>250	Fresh basalt, chert, diabase, gneiss, granite, quartzite

Source: Hoek, E. and Brown, E.T., *Int. J. Rock Mech. Min. Sci. Geomech. Abstr.*, 34, 1165, 1997.

Source: ISRM, *Int. J. Rock Mech. Min. Sci. Geomech. Abstr.*, 15, 319, 1978.

[a] Slightly different to classification in geotechnical context.

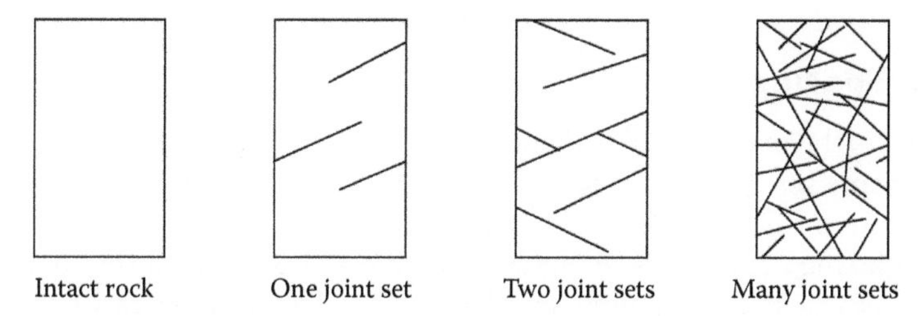

Figure 4.3 Number of joint sets.

the rock types that fall into each group and simple index tests that can be carried out in the field to classify them.

The rock mass can have any number of joints. When there are no joints, ideally, the rock mass and the intact rock should have the same properties, provided the rock is homogeneous. Joints within a joint set are approximately parallel. One can even define an average spacing for a joint set. This is simply the distance between the two adjacent joints in the same set. An increasing number of joints and joint sets make the rock mass more and more fragmented, thus increasing the degrees of freedom of the individual pieces. In addition, the block sizes become smaller. This is evident in Figure 4.3, showing a diagrammatic representation of intact rock and rock masses with one or more joint sets.

4.3 FACTORS AFFECTING DISCONTINUITIES

There are several parameters that are used to describe discontinuities and the rock mass. They are

- Orientation
- Spacing
- Persistence
- Roughness
- Wall strength
- Aperture
- Filling
- Seepage
- Number of joint sets
- Block size and shape

4.3.1 Orientation

Orientation of the discontinuity, measured by the dip and dip direction, is very critical to stability, as we observed during the discussion of kinematic analysis in Chapter 2. By locating and/or aligning the structure (e.g. tunnel) in the right direction, the stability can be improved significantly.

4.3.2 Spacing

Spacing is the perpendicular distance between two adjacent discontinuities of the same set. It affects the hydraulic conductivity of the rock mass and the failure mechanism. Closely spaced joints can imply highly permeable rock. Spacing determines the intact rock block sizes within the rock mass, with closer spacing implying smaller blocks. The spacing can be used to describe the rock mass as shown in Table 4.2.

4.3.3 Persistence

Persistence is a measure of the extent to which the discontinuity extends into the rock. In other words, what is the surface area of the discontinuity? This is the area that takes part in any possible sliding, and hence is an important parameter in determining stability. Although this is an important parameter in characterising the rock mass, it is quite difficult to determine. The trace length of the discontinuity, on the exposed surface, is often taken as a crude measure of the persistence. Persistence of a rock mass can be described on the basis of Table 4.3. Spacing and persistence are two parameters that control the sizes of the blocks of intact rocks that make up the rock mass. They are both measured by a measuring tape.

Table 4.2 Rock classification based on the spacing of discontinuities

Description	*Spacing (mm)*
Extremely close spacing	<20
Very close spacing	20–60
Close spacing	60–200
Moderate spacing	200–600
Wide spacing	600–2000
Very wide spacing	2000–6000
Extremely wide spacing	>6000

Source: ISRM, *Int. J. Rock Mech. Min. Sci. Geomech. Abstr.*, 15, 319, 1978.

Table 4.3 Description for persistence

Description	*Trace length (m)*
Very low persistence	<1
Low persistence	1–3
Medium persistence	3–10
High persistence	10–20
Very high persistence	>20

Source: ISRM, *Int. J. Rock Mech. Min. Sci. Geomech. Abstr.*, 15, 319, 1978.

4.3.4 Roughness

The roughness of a rock joint refers to the large-scale surface undulations (waviness) observed over several metres and the small-scale unevenness of the two sides relative to the mean plane, observed over several centimetres (see Figure 4.4). The large-scale undulations can be called stepped, undulating or planar; the small-scale unevenness can be called rough, smooth or slickensided. Figure 4.4 shows the three major large-scale undulations. Close-up views in Figure 4.4 show two (rough and smooth) of the three small-scale unevenness profiles. Slickenside is a standard term used for smooth and slick, shiny surfaces that look polished. Combining the large-scale undulations and small-scale unevenness, the roughness of a joint can be classified as shown in Table 4.4, where the roughness decreases (in a broad sense) from Class I to IX. Large-scale surface undulations have a greater influence on the roughness than the small-scale unevenness, and this is reflected in Table 4.4. Although it is clear that when it comes to roughness, I > II > III, IV > V > VI, VII > VIII > IX, and I > IV > VII, II > V > VIII, III > VI > IX, it is not always the case that class III is rougher than VII.

Roughness is an important factor governing the shear strength of the joint, especially when the discontinuity is undisplaced or interlocked. When displaced or the joints are infilled, the interlock is lost, and the roughness is less important. Under such circumstances, the shear strength characteristics of

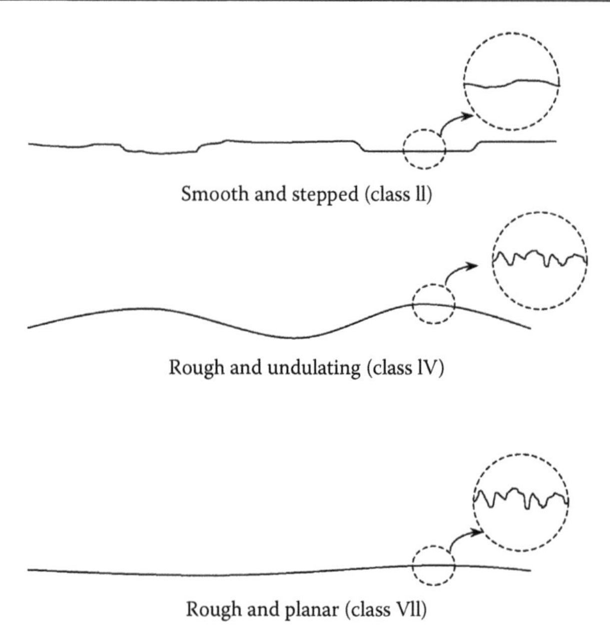

Figure 4.4 Three different roughness profiles.

Table 4.4 Roughness classification

Class	*Unevenness and undulations*	J_r	
I	Rough, stepped	4[a]	Increasing roughness⇒
II	Smooth, stepped	3[a]	
III	Slickensided, stepped	2[a]	
IV	Rough, undulating	3	
V	Smooth, undulating	2	
VI	Slickensided, undulating	1.5	
VII	Rough, planar	1.5	
VIII	Smooth, planar	1	
IX	Slickensided, planar	0.5	

Source: ISRM, *Int. J. Rock Mech. Min. Sci. Geomech. Abstr.*, 15, 319, 1978.

Note: Slickenside = polished and striated surface.

[a] J_r values for I, II and III as suggested by Barton (1987) and others by Hoek et al. (2005).

the infill material govern the shear strength along the joint. Figure 4.4 shows three of the possible nine roughness profiles suggested in Table 4.4. The large-scale undulations and small-scale unevenness are shown separately. The joint roughness number J_r given in the table is used later in rock mass classification using the *Q*-system, which is discussed in more detail in Section 4.6.

There are special techniques such as the linear profiling method, compass and disc-clinometer method and photogrammetric method available for measuring roughness. It is measured along the dip direction. Barton (1973) defined the term joint roughness coefficient (JRC), a value ranging from 0 for smooth or slickensided planar surfaces to 20 for rough stepped or undulating surfaces. Roughness profiles with corresponding JRC values as suggested by Barton and Choubey (1977) are also reproduced in ISRM (1978). JRC can be estimated visually by comparing the surface profiles with these standard ones.

4.3.5 Wall strength

Wall strength refers to the compressive strength of the rock that makes up the walls of the discontinuity. Barton (1973) introduced the term joint wall compressive strength (JCS) to describe wall strength, which was later refined by Barton and Choubey (1977). This is an important factor that governs the shear strength and deformability. In unaltered joints, the uniaxial compressive strength (UCS) can be taken as JCS. When the joint surface is weathered, JCS can conservatively (i.e. lower bound) be taken as 25% of UCS.

The point load test or Schmidt hammer test are other possibilities here that can be used for estimating the UCS. JCS can be determined from the Schmidt hammer rebound number as follows (Franklin and Dusseault, 1989):

$$\log_{10} \text{JCS (MPa)} = 0.00088\gamma R + 1.01 \tag{4.1}$$

where γ = unit weight of the rock (kN/m^3) and R = Schmidt hammer rebound number on the joint surface.

The peak friction angle ϕ_{peak} of an unfilled joint can be in the range of 30–70°. When the joint walls are not weathered, the residual friction angle ϕ_r is typically in the range of 25–35°. In the case of weathered joint walls, ϕ_r can be as low as 15°.

The friction angle of a rough discontinuity surface has two components: basic friction angle of the rock material ϕ_b, and the roughness angle due to interlocking of the surface irregularities or asperities i. Therefore, when cohesion is neglected, the shear strength can be written as (remember the Mohr–Coulomb failure criterion from soil mechanics):

$$\tau = \sigma_n \tan(\phi_b + i) \tag{4.2}$$

where σ_n is the effective normal stress on the discontinuity plane. The basic friction angle ϕ_b is approximately equal to the residual friction angle ϕ_r. The roughness angle i (in degrees) can be estimated by

$$i = \text{JRC} \log\left(\frac{\text{JCS}}{\sigma_n}\right) \tag{4.3}$$

At low values of effective normal stresses, the roughness angle estimated from Equation 4.3 can be unrealistically large. For designs, it is suggested that $\phi_b + i$ should be limited to 50° and JCS/σ_n should be in the range of 3–100 (Wyllie and Mah, 2004). Substituting Equation 4.3 into Equation 4.2, we can express the shear strength as

$$\tau = \sigma_n \tan\left[\phi_b + \text{JRC}\log\left(\frac{\text{JCS}}{\sigma_n}\right)\right] \quad (4.4)$$

An average value of ϕ_b can be taken as 30° (ISRM, 1978). The roughness angle i can be as high as 40°. At the very early stages of movement along the discontinuity planes, there is relatively high interlocking due to the surface roughness, with a friction angle of $\phi + i$. When the asperities are sheared off, the roughness angle i decreases to zero, and the friction angle reaches the residual friction angle. In Equation 4.4, ϕ_b can be replaced by ϕ_r.

4.3.6 Aperture

A discontinuity can be closed, open or filled. Aperture is the perpendicular distance between the two adjacent rock walls of an open discontinuity (Figure 4.2b), where the space is filled by air or water. The joint is called tight or open, depending on whether the aperture is small or large. Aperture is generally greater near the surface due to stress relief, and becomes less with depth. Apertures can be described using terms given in Table 4.5. When the space between the walls is filled (Figure 4.2a) with sediments, we will not use the term aperture – we call it width of the infill. Measuring tape or a feeler gauge can be used for measuring aperture.

Table 4.5 Descriptions associated with apertures

Aperture (mm)		*Description*
<0.1	Very tight	Closed features
0.1–0.25	Tight	
0.25–0.5	Partly open	
0.5–2.5	Open	Gapped features
2.5–10	Moderately wide	
>10	Wide	
10–100	Very wide	Open features
100–1000	Extremely wide	
>1000	Cavernous	

Source: ISRM, *Int. J. Rock Mech. Min. Sci. Geomech. Abstr.*, 15, 319, 1978.

4.3.7 Filling

Filling is the term used to describe the material (e.g. calcite, chlorite, clay and silt) that occupies the space between the adjacent rock walls of a discontinuity (Figure 4.2a). Its properties can differ significantly from those of the rocks on either side. It affects the permeability and the deformability of the rock mass. A complete description of the filling may include the width, mineralogy, grain size, water content, permeability and strength (see Table 4.1). Depending on the nature of the project, relevant laboratory tests may be carried out on the fillings to assess their characteristics.

4.3.8 Seepage

In a rock mass, seepage occurs mainly through discontinuities (secondary permeability), as the permeability of the intact rock (primary permeability) is generally very low. The observation is generally visual and hence subjective. An excavation can range from being literally dry to one that has heavy inflow of water. ISRM (1978) has separate ratings from I (no seepage) to VI (heavy flow) for unfilled and filled discontinuities. It also gives ratings from I (no seepage) to V (exceptionally high inflow) for tunnel walls on the basis of seepage. The presence of water can reduce the shear strength along the joint and can have adverse effects on the stability.

4.3.9 Number of joint sets

The number of joint sets in the system of discontinuities is one of the factors used in classifying the rock mass. It determines the ability of the rock mass to deform without actually undergoing any failure within the intact rock. As the number of joint sets increases, the individual block size decreases and their degrees of freedom to move increase. The rock mass can be classified based on number of joint sets as given in Table 4.6. About 100–150

Table 4.6 Classification based on number of joint sets

Group	*Joint sets*
I	Massive, occasional random joints
II	One joint set
III	One joint set plus random
IV	Two joint sets
V	Two joint sets plus random
VI	Three joint sets
VII	Three joint sets plus random
VIII	Four or more joint sets
IX	Crushed rock, similar to soils

Source: ISRM, *Int. J. Rock Mech. Min. Sci. Geomech. Abstr.*, 15, 319, 1978.

joints must be located, and their dip and dip directions be measured for generating a pole plot (see Chapter 2). These can be used to identify the number of joint sets present.

4.3.10 Block size

The rock mass consists of blocks formed by intersections of several joints. Block size in a rock mass depends on the number of discontinuity sets, spacing and persistence that separates the blocks. It is similar to grain size is soils. The blocks can be in the form of cubes, tetrahedrons, sheets and so on. The block size and the interblock shear strength at the face of the discontinuities play a key role in the stability of the rock mass in rock slopes and underground openings. It is a key parameter in rock mass classification.

Block size is defined as the average diameter of an equivalent sphere of the same volume. It is quantified by block size index I_b, the average dimension of a typical block, or volumetric joint count J_v, the total number of joints intersecting a unit volume of rock mass. Rock quality designation (RQD) also is a measure of the block size – the larger the RQD, the larger the blocks. In the case of an orthogonal joint system of three sets with spacing of S_1, S_2 and S_3, the block size index is defined as follows:

$$I_b = \frac{S_1 + S_2 + S_3}{3} \tag{4.5}$$

There are $1/S_1$, $1/S_2$ and $1/S_3$ joints per metre along the three orthogonal directions, where S_1, S_2 and S_3 are in metres. The volumetric joint count (in joints/m^3) is defined as the sum of the number of joints per metre for each joint set present, and is given as follows:

$$J_v = \frac{1}{S_1} + \frac{1}{S_2} + \cdots + \frac{1}{S_n} \tag{4.6}$$

ISRM (1978) suggests that RQD and J_v can be related by

$$\text{RQD} = 115 - 3.3J_v \tag{4.7}$$

For $J_v < 4.5$, RQD is taken as 100% and for $J_v > 30$, RQD is taken as 0%. ISRM (1978) suggests some standard descriptions for the block sizes based on J_v (Table 4.7).

EXAMPLE 4.1

A rock mass consists of four joint sets. The following joint counts are made normal to each set: joint set 1 = 12 per 10 m, joint set 2 = 17 per 5 m, joint set 3 = 16 per 5 m and joint set 4 = 13 per 10 m. Find the volumetric joint count. How would you describe the block size?

Estimate the RQD.

Solution

The joint spacings are given by $S_1 = 10/12$ m, $S_2 = 5/17$ m, $S_3 = 5/16$ m and $S_4 = 10/13$ m. Applying Equation 4.6, we get

$$J_v = \frac{12}{10} + \frac{17}{5} + \frac{16}{5} + \frac{13}{10} = 9.1 \text{ joints per m}^3 \rightarrow \text{Medium-sized blocks}$$

From Equation 4.7, RQD = 115 – (3.3 × 9.1) = 85

The rock mass is described by one of the following adjectives reflecting the block size and shape (ISRM, 1978).

- Massive – few joints or very wide spacing
- Blocky – approximately equidimensional
- Tabular – one dimension considerably smaller than the other two
- Columnar – one dimension considerably larger than the other two
- Irregular – wide variations of block size and shape
- Crushed – heavily jointed to sugar cubes

The common methods of measurements of the 10 parameters listed at the beginning of this section and their relative merits are summarised in Table 4.8. The relative presence of these parameters is illustrated in Figure 4.5.

4.4 ROCK MASS CLASSIFICATION

In soils, we have been using different soil classification systems, such as the Unified Soil Classification System (USCS), American Association of State Highway Transportation Officials (AASHTO), British Standards (BS) and Australian Standards (AS). The main objective has been to group soils

Table 4.7 Block sizes and J_v values

J_v *(Joints/m³)*	*Description*
<1	Very large blocks
1–3	Large blocks
3–10	Medium-sized blocks
10–30	Small blocks
30–60	Very small blocks
>60	Crushed rock

Source: Franklin, J.A. and Dusseault, M.B., *Rock Engineering*, McGraw Hill, New York, p. 600, 1989.

Source: ISRM, *Int. J. Rock Mech. Min. Sci. Geomech. Abstr.*, 15, 319, 1978.

Table 4.8 Methods of measurements of discontinuity parameters

Parameter	*Method of measurement*	*Core*	*Borehole via TV camera*	*Exposure*
Orientation	Compass-inclinometer	M	G	G
Spacing	Measuring tape	G	G	G
Persistence	Measuring tape	B	B	G/M
Roughness	Against reference chart	M	B	G
Wall strength	Schmidt hammer	M	B	G
Aperture	Scale or feeler gauge	B	M	G
Filling	Visual	B	B	G
Seepage	Timed observations	B	B/M	G
Number of sets	Stereographic projections	M	G	G
Block size	3D fracture frequency	B	B	G

G = Good, M = Medium, B = Bad.

Source: Hudson, J.A., *Rock Mechanics Principles in Engineering Practice*, Butterworths, London, 1989.

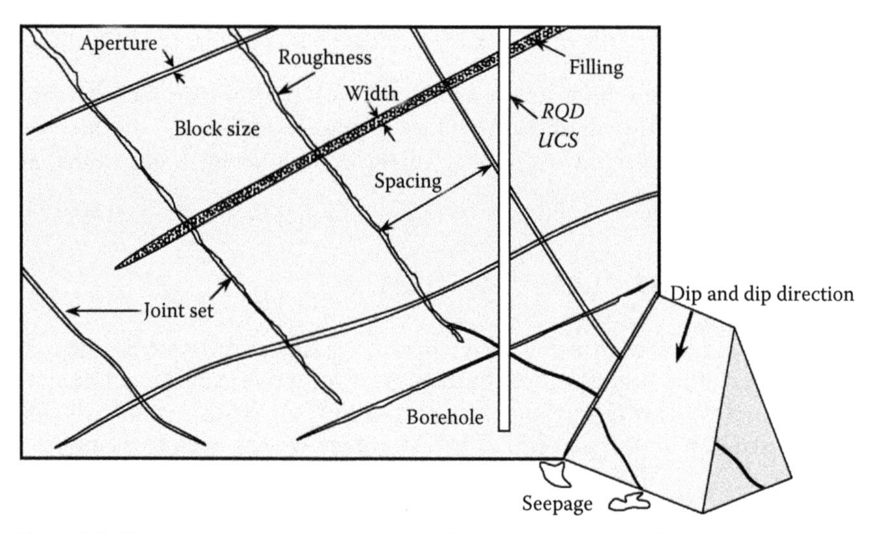

Figure 4.5 Diagrammatic representations of parameters describing discontinuities. (Adapted from Hudson, J.A., *Rock Mechanics Principles in Engineering Practice*, Butterworths, London, 1989.)

of similar behaviour and to develop some systematic ways to describe soils without any ambiguity. This is not any different with rocks.

A rock mass is classified on the basis of three factors: (1) intact rock properties, (2) joint characteristics and (3) boundary conditions. When it comes to intact rock properties, the strength and stiffness (Young's modulus) are the two parameters that are used in designs. In rock mass classification, it is the UCS that is commonly used as a measure of strength. The stability of the jointed rock mass is severely influenced by the frictional resistance along the joint between the adjacent blocks. The joint surface can be stepped or undulated (macroscopically) and very rough at the contact points, implying very high shear strength. However, when the joints are filled, the aperture width and the characteristics of the filling become more important than the characteristics of the rock wall roughness. The third factor is the boundary conditions, which include the in situ stresses present within the rock mass and the groundwater conditions. Groundwater has adverse effects on the stability by increasing the pore water pressure; it reduces the effective stress and therefore reduces the shear strength.

With a wide range of strength values for the intact rock cores and so many different parameters to describe the discontinuities and the rock mass, there is certainly a need to have some classification systems for rocks too. The classification systems ensure that we all speak the same language when referring to a specific rock mass. Some of the common rock mass classification systems are as follows:

- Rock mass rating (RMR)
- *Q*-system
- Geological strength index (GSI)

These are discussed in detail in the following sections. They are commonly used for designing the underground openings such as tunnels and excavations.

4.5 ROCK MASS RATING

Rock mass rating (RMR), also known as the Geomechanics Classification System, was originally proposed by Bieniawski in 1973 at the Council for Scientific and Industrial Research in South Africa. It was slightly modified in 1989, based on the analysis of data from 268 tunnel sites in hard rock areas. It is a rating out of a maximum of one hundred, based on the first five parameters listed below:

- Strength of intact rock (Table 4.9) – maximum score of 15
- RQD (Table 4.10) – maximum score of 20
- Mean spacing of the discontinuities (Table 4.11) – maximum score of 20
- Condition of discontinuities (Table 4.12) – maximum score of 30
- Groundwater conditions (Table 4.14) – maximum score of 15
- Orientation of discontinuities (Table 4.15)

The ratings of the first five factors are added to make up the RMR, which lies in the range of 0–100. The last one is an adjustment to the RMR considering how favourable or unfavourable the joint orientations are with respect to the project. These values are negative, from 0 to −60, and are different for tunnels, foundations and slopes.

The strength of intact rock can be quantified by UCS or the point load strength index $I_{s(50)}$. The corresponding rating increments are given in Table 4.9. Hoek and Brown (1997) noted that point load tests are unreliable when UCS is less than 25 MPa. For weaker rocks, it is recommended that the point load strength index is not used when assigning ratings for classification. Deere and Miller's (1966) strength classification was used as the basis in assigning these rating increments, and UCS and $I_{s(50)}$ values in Table 4.9 are related by UCS = $25I_{s(50)}$.

The rating increments for the drill core quality (represented by RQD) are given in Table 4.10. RQD can vary depending on the direction of the borehole.

The rating increments based on the mean spacing of discontinuities are given in Table 4.11. Very often, there are more than one set of discontinuities present within the rock mass. The set of discontinuities that is the most critical for the specific project must be considered in assigning the rating increment. The wider the joint spacing, the lesser is the deformation within

Table 4.9 Rating increments for uniaxial compressive strength

Point load strength index, $I_{s(50)}$ (MPa)	*UCS (MPa)*	*Rating*
Not applicable; use UCS only	<1	0
	1–5	1
	5–25	2
1–2	25–50	4
2–4	50–100	7
4–10	100–250	12
>10	>250	15

Source: Bieniawski, Z.T., *Engineering Rock Mass Classification*, Wiley Interscience, New York, p. 251, 1989.

Table 4.10 Rating increments for RQD

RQD (%)	<25	25–50	50–75	75–90	90–100
Rating	3	8	13	17	20

Source: Bieniawski, Z.T., *Engineering Rock Mass Classification*, Wiley Interscience, New York, p. 251, 1989.

Table 4.11 Rating increments for joint spacing

Spacing (mm)	<60	60–200	200–600	600–2000	>2000
Rating	5	8	10	15	20

Source: Bieniawski, Z.T., *Engineering Rock Mass Classification*, Wiley Interscience, New York, p. 251, 1989.

the rock mass, and hence the higher the rating increments. When there are joint sets with spacing of S_1, S_2, S_3 and so on, the average spacing can be computed as follows:

$$\frac{1}{S_{avg}} = \frac{1}{S_1} + \frac{1}{S_2} + \frac{1}{S_3} \cdots \tag{4.8}$$

Hudson and Priest (1979) analysed 7000 joint spacing values measured in chalk at Chinnor tunnel in England and proposed the following relationship between RQD and the mean joint frequency λ per unit length (m):

$$\text{RQD} = 100e^{-0.1\lambda}\left(0.1\lambda + 1\right) \tag{4.9}$$

where λ is the number of joints per metre. In the absence of measurements of joint spacing, Equation 4.9 can be used to estimate the joint frequency and thus joint spacing.

EXAMPLE 4.2

Estimate the joint spacing of a rock mass where RQD = 81%.

Solution

From Equation 4.9, for RQD = 81%, λ = 8 per m.

Therefore, the joint spacing = 1/8 m = 0.125 m = 125 mm.

The rating increments for the condition of the discontinuities are given in Table 4.12. Here too, the joint set that is the most critical to the project should be considered in assigning the rating. In general, the weakest and smoothest joint set should be considered.

Table 4.12 Rating increments for the joint condition

Condition of joint	*Rating*
Open joint infilled with soft gouge >5 mm thickness OR separation >5 mm, and continuous extending several metres	0
Smooth surfaces OR 1–5 mm gouge infilling OR 1–5 mm aperture, and continuous joint extending several metres	10
Slightly rough surfaces, aperture <1 mm, and highly weathered walls	20
Slightly rough surfaces, <1 mm separation, slightly weathered walls	25
Very rough surfaces, not continuous joints, no separation, unweathered wall	30

Source: Bieniawski, Z.T., *Engineering Rock Mass Classification*, Wiley Interscience, New York, p. 251, 1989.

Table 4.13 Guidelines for classifying the condition of discontinuity

Persistence (m)	<1	1–3	3–10	10–20	>20
Rating	6	4	2	1	0
Aperture (mm)	None	<0.1	0.1–1.0	1–5	>5
Rating	6	5	4	1	0
Roughness	Very rough	Rough	Slightly rough	Smooth	Slickensided
Rating	6	5	3	1	0
Infilling (gouge)	None	Hard filling <5 mm	Hard filling >5 mm	Soft filling <5 mm	Soft filling >5 mm
Rating	6	4	2	2	0
Weathering	Unweathered	Slightly weathered	Moderately weathered	Highly weathered	Decomposed
Rating	6	5	3	1	0

Source: Bieniawski, Z.T., *Engineering Rock Mass Classification*, Wiley Interscience, New York, p. 251, 1989.

Source: Hoek, E., et al., *Support of Underground Excavations in Hard Rock*, A.A. Balkema, Rotterdam, 2005.

Gouge is a fine filling material between the joint walls that is formed by the grinding action between the two walls. It can be in the form of silt, clay, rock flour and the like, and can be a few centimetres in thickness. Table 4.12 is adequate when there is little information about the joints. In the presence of more detailed information about the joint, the guidelines in Table 4.13 can be used for a more thorough classification of the joint conditions.

EXAMPLE 4.3

A joint with slightly rough and weathered walls has a separation less than 1 mm. What would be the rating increment for the joint conditions?

Solution

From Table 4.12, the rating increment is 25.

EXAMPLE 4.4

In Example 4.3, if the following information is available, how would you modify the rating increment for the joint condition?

Persistence = 2 m, aperture = 0.1–0.5 mm, roughness = slightly rough, infilling = none, weathering = slight

Solution

From Table 4.13, the rating increment = 4 + 4 + 3 + 6 + 5 = 22.

The presence of groundwater in the joints can severely influence the shear strength and the deformability of the rock mass. The rating increment for the groundwater conditions is based on (1) inflow (L/min) per 10 m of tunnel length, (2) ratio of joint water pressure to major principal stress or (3) the general wetness condition of the joint. The general condition (e.g. damp) of the joint can be determined qualitatively from the drill cores and bore logs, in the absence of exploratory audits or pilot tunnels. These rating increments are given in Table 4.14.

Although it is not possible to do much about the intact rock strength, discontinuities, and the groundwater conditions in the rock mass, it is certainly possible to improve the stability of the proposed structure by orienting it in the best possible way. Bieniawski (1989) assigned negative rating increments depending on how favourable or unfavourable the orientations of the discontinuities are with respect to the project. These rating increments, given in Table 4.15, often called rating adjustments, are different for tunnels, foundations and slopes.

Rating adjustments in Table 4.15 are rather subjective. It requires some sound judgement in assigning the rating adjustments for the discontinuity orientations. Consultation with an engineering geologist familiar with the rock formation and the project is very valuable here.

Table 4.14 Rating increments for groundwater conditions

Inflow (L/min) per 10-m tunnel length	*Joint water pressure/major principal stress*	*General conditions*	*Rating increment*
>125	>0.5	Flowing	0
25–125	0.2–0.5	Dripping	4
10–25	0.1–0.2	Wet	7
<10	<0.1	Damp	10
None	0	Completely dry	15

Source: Bieniawski, Z.T., *Engineering Rock Mass Classification*, Wiley Interscience, New York, p. 251, 1989.

Table 4.15 Rating adjustments for discontinuity orientations

Orientation of joints with respect to the project	*Rating increments*		
	Tunnels and mines	*Foundations*	*Slopes*
Very unfavourable	−12	−25	−60
Unfavourable	−10	−15	−50
Fair	−5	−7	−25
Favourable	−2	−2	−5
Very favourable	0	0	0

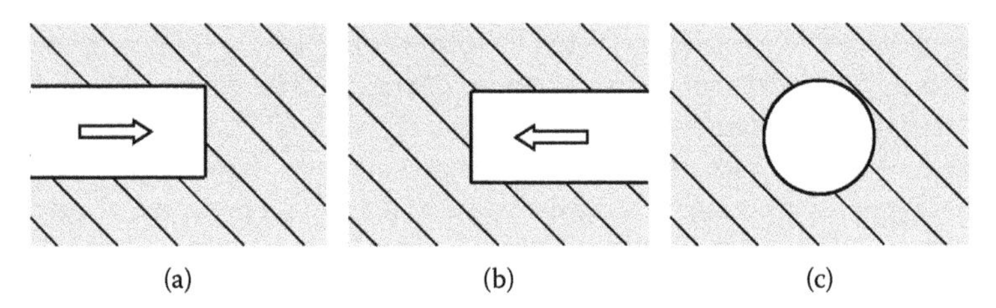

Figure 4.6 Tunnelling: (a) drive with dip, (b) drive against dip and (c) tunnel axis parallel to strike.

Table 4.16 Effects of discontinuity orientation in tunnelling

Strike perpendicular to tunnel axis				*Strike parallel to tunnel axis*		*Dip 0–20° irrespective of strike*
Drive with dip		*Drive against dip*				
Dip 45–90°	Dip 20–45°	Dip 45–90°	Dip 20–45°	Dip 45–90°	Dip 20–45°	
Very favourable	Favourable	Fair	Unfavourable	Very unfavourable	Fair	Fair

Let us consider some tunnelling work. The strike of the discontinuity plane can be perpendicular (Figure 4.6a and b) or parallel (Figure 4.6c) to the tunnel axis. When it is perpendicular, depending on whether the tunnel is driven with (Figure 4.6a) or against (Figure 4.6b) the dip, the rating adjustments should be different. Some guidelines for choosing the appropriate adjective in the first column of Table 4.15 in tunnelling work are given in Table 4.16.

Adding up all five empirical rating increments and the negative rating adjustments for orientations, a total score out of 100 is obtained. This is known as the RMR value.

EXAMPLE 4.5

Determine the RMR value for tunnelling work in a rock formation with the following details.

- The point load strength index $I_{s(50)} = 6$ MPa
- RQD = 80%
- Mean spacing of discontinuities = 500 mm
- Joint walls had slightly rough and weathered surfaces, with less than 1-mm separation
- Groundwater condition = Damp
- Discontinuity orientation with respect to the project = Fair

Solution

From Tables 4.9, 4.10, 4.11, 4.12 and 4.14, the score = 12 + 17 + 10 + 25 + 10 = 74. Taking the discontinuity orientation into consideration (Table 4.15), with the rating adjustment of –5, the RMR becomes 69.

Based on the RMR value, including the adjustment for the discontinuity orientation, a rock mass can be classified and described as given in Table 4.17 (Bieniawski, 1989).

Tunnelling is common in mining engineering when accessing the mineral deposits from deep inside the Earth. Tunnels are also used in transportation by trains and vehicles. Further, tunnels are used to carry water, sewage and gas lines over long distances. In tunnelling work, stand-up time is the time that an open excavation can stand unsupported before it caves in. Of course, it depends on the length of the tunnel. This is an important consideration in tunnelling work. The approximate relationship between the rock mass class (Table 4.17) and stand-up time in a tunnel, along with the cohesion and friction angle of the rock mass, is given in Table 4.18.

Bieniawski (1989) suggested guidelines for selecting the excavation and support procedures (e.g. rock bolt and shotcrete) for underground openings such as tunnels on the basis of the rock mass class, derived from RMR. Noting the fact that RMR was originally developed for tunnelling, based on civil engineering case studies, Laubscher (1977) extended this to mining as mining rock mass rating (MRMR). The MRMR has further adjustments for in situ and mining-induced stresses, effects of blasting and weathering of the parent rock.

Table 4.17 Rock mass classes based on RMR

RMR	81–100	61–80	41–60	21–40	0–20
Class number	I	II	III	IV	V
Description	Very good rock	Good rock	Fair rock	Poor rock	Very poor rock

Table 4.18 Meaning of rock mass class

Class number	*I*	*II*	*III*	*IV*	*V*
Average stand-up time of tunnel	20 years for 15-m span	1 year for 10-m span	1 week for 5-m span	10 hours for 2.5-m span	30 minutes for 1-m span
Cohesion of rock mass (kPa)	>400	300–400	200–300	100–200	<100
Friction angle of rock mass (°)	>45	35–45	25–35	15–25	<15

Source: Bieniawski, Z.T., *Engineering Rock Mass Classification*, Wiley Interscience, New York, p. 251, 1989.

4.6 TUNNELLING QUALITY INDEX: Q-SYSTEM

Barton et al. (1974) of the Norwegian Geotechnical Institute proposed the Tunnelling Quality Index, known as Q, a new rock mass classification system. The system was developed based on several case histories, and the objective was to characterise the rock mass and determine the tunnel support requirements. Similar to RMR, the Tunnelling Quality Index Q is derived based on six parameters listed below:

- RQD (0–100)
- Joint set number, J_n (1–20)
- Joint roughness number, J_r (1–4)
- Joint alteration number, J_a (1–20)
- Joint water reduction factor, J_w (0.1–1.0)
- Stress reduction factor, SRF (1–20)

It is defined as follows:

$$Q = \left(\frac{\text{RQD}}{J_n}\right)\left(\frac{J_r}{J_a}\right)\left(\frac{J_w}{\text{SRF}}\right) \tag{4.10}$$

The numerical value of Q ranges on a logarithmic scale from 0.001 to 1000+, covering the whole spectrum of rock mass from a heavily jointed weak rock mass to sound unjointed rock. The higher the value of Q, the better the rock mass quality. The three numerators in the quotients, RQD, J_r and J_w, are assigned values such that their higher values reflect better quality rock mass. The three denominators, J_n, J_a and SRF, are assigned values such that their lower values reflect better quality rock mass. Barton (2002) suggested slight modifications to the original Q-system, particularly to J_a and SRF.

RQD and J_n are both reflections of the number of joints present within the rock mass. The higher the RQD, the lower the J_n, and vice versa. As a result, the first quotient RQD/J_n in Equation 4.10 can take a wide range of values from 0.5 to 200+. These values are seen crudely as the block sizes in centimetres (Barton et al., 1974). The RQD values and rock classifications (Table 4.19) are quite similar to those used in RMR classification. The joint set number J_n is assigned on the basis of Table 4.20. J_n is 1.0 for rock

Table 4.19 RQD values in Q-system

Class	*A*	*B*	*C*	*D*	*E*
Designation	Very poor	Poor	Fair	Good	Excellent
RQD	0–25	25–50	50–75	75–90	90–100

Source: Barton, N.R., et al., *Rock Mech.*, 6, 189, 1974.

Notes: (1) When RQD <10, use 10 in computing Q; (2) RQD rounded off to 5 (i.e. 80 and 85) is adequate.

with no joints and is assigned the maximum possible value of 20 when it is crushed. J_n increases with increasing number of joint sets, reflecting lower values of Q in Equation 4.10.

The second quotient, J_r/J_a, in Equation 4.10 is a measure of shear strength. The joint roughness number J_r is a measure of the joint roughness and lies in the range of 0.5–4, with larger numbers representing rougher joints, implying greater shear strength. Rocks with discontinuous joints (i.e. low persistence) are assigned the maximum value of 4, and those with continuous slickensided planar joints are assigned the minimum value of 0.5. Suggested values of J_r are given in Table 4.21. It can

Table 4.20 Joint set number J_n for Q-system

Class	*Description*	J_n
A	Massive; none or few joints	0.5–1.0
B	One joint set	2
C	One joint set plus random joints	3
D	Two joint sets	4
E	Two joint sets plus random joints	6
F	Three joint sets	9
G	Three joint sets plus random joints	12
H	Four or more joint sets; random; heavily jointed; sugar cubes etc.	15
J	Crushed rock; earth-like	20

Source: Barton, N.R., et al., *Rock Mech.*, 6, 189, 1974.

Notes: (1) For tunnel intersections, use $3.0 \times J_n$; (2) for portals, use $2.0 \times J_n$.

Table 4.21 Joint roughness number J_r

Class	*Description*	J_r
(a) Rock–wall contact, and (b) Rock–wall contact before 10 cm of shear		
A	Discontinuous joints	4
B	Rough or irregular, undulating	3
C	Smooth, undulating	2
D	Slickensided, undulating	1.5
E	Rough or irregular, planar	1.5
F	Smooth, planar	1.0
G	Slickensided, planar	0.5
(c) No rock–wall contact when sheared		
H	Zone containing clayey minerals thick enough to prevent rock–wall contact	1.0
J	Sandy, gravelly or crushed zone thick enough to prevent rock–wall contact	1.0

Source: Barton, N.R., et al., *Rock Mech.*, 6, 189, 1974.

Notes: (1) Add 1.0 if the mean spacing of the relevant joint set is greater than 3 m and (2) $J_r = 0.5$ for planar slickensided joints having lineations if the lineations are favourably oriented.

be seen that Table 4.21 correlates with the ISRM-suggested roughness classes given in Table 4.4.

Joint alteration number, J_a, is a measure of the degree of alteration of the joint wall or infill material, which is quantified in terms of residual friction angle ϕ_r. Tan^{-1} (J_r/J_a) is a fair approximation of the residual friction angle. Noting that there is no cohesion at the residual state, the residual shear strength is given by $\tau \approx \sigma_n (J_r/J_a)$. The weakest joint set (i.e. with the lowest J_r/J_a value) with due consideration to the orientation with respect to stability should be used in computing the Q-value. The suggested values of J_a are given in Table 4.22. It can be seen that the rough and unaltered joint sets get larger values of J_r/J_a than the smooth slickensided joints with clay fillings. Surface staining on rocks occurs due to moisture and presence of chemicals.

Rough and unaltered joints that are in direct contact have undergone very low strains, and hence the shear strength along such a surface is

Table 4.22 Joint alteration number J_a

Class	*Description*	ϕ_r (°)	J_a
(a) Rock–wall contact (no mineral fillings, only coatings)			
A	Tightly healed, hard, nonsoftening, impermeable filling, i.e. quartz or epidote		0.75
B	Unaltered joint walls, surface staining only	25–35	1.0
C	Slightly altered joint walls, nonsoftening mineral coatings, sandy particles, clay-free disintegrated rock etc.	25–30	2.0
D	Silty- or sandy-clay coatings, small clay fraction (nonsoftening)	20–25	3.0
E	Softening or low-friction clay mineral coatings, i.e. kaolinite or mica; also chlorite, talc, gypsum, graphite etc., and small quantities of swelling clays	8–16	4.0
(b) Rock–wall contact before 10-cm shear (thin mineral fillings)			
F	Sandy particles, clay-free disintegrated rock etc.	25–30	4.0
G	Strongly overconsolidated nonsoftening clay material fillings (continuous but <5-mm thickness)	16–24	6.0
H	Medium or low overconsolidation, softening clay mineral fillings (continuous but <5-mm thickness)	12–16	8.0
J	Swelling clay fillings, i.e. montmorillonite (continuous but <5-mm thickness); value of J_a depends on % of swelling clay-size particles and access to water etc.	6–12	8–12

Table 4.22 Joint alteration number J_a (*Continued*)

Class	*Description*	ϕ_r (°)	J_a
(c) No rock–wall contact when sheared (thick mineral fillings)			
K, L, M	Zones or bands of disintegrated crushed rock and clay (see G, H and J for description of clay condition)	6–24	6, 8 or 8–12
N	Zones or bands of silty- or sandy-clay, small clay fraction (nonsoftening)	—	5.0
O, P, R	Thick continuous zones or bands of clay (see G, H and J for description of clay condition)	6–24	10, 13 or 13–20

Source: Barton, N., *Int. J. Rock Mech. Min. Sci.*, 39, 185, 2002.

Source: Barton, N.R., et al., *Rock Mech.*, 6, 189, 1974.

Note: ϕ_r values are approximate.

closer to the peak value than the residual values. These surfaces will dilate when sheared, which favours the stability of the tunnels. When the joints are filled or have thin mineral coatings, the shear strength would be significantly lower. In some situations where the mineral filling is rather thin, the rock–wall contact takes place after some shear (case b in Table 4.22), which minimises further slide. When the filling is thick, there will be no contact even after some shear (case c in Table 4.22), enabling the residual strength to be reached. Such situations are unfavourable in tunnelling work.

The third quotient in Equation 4.10, J_w/SRF, is something Barton et al. (1974) referred to as an 'active stress' term. It is well known that water can reduce the effective normal stress (σ'), which in turn reduces the shear strength. Further, water can soften and possibly wash out the infill. The joint water reduction factor J_w, which ranges from 0.05 to 1.0, accounts for such reduction in shear strength due to the presence of water in the rock mass. A dry excavation is assigned a factor of 1.0 and a situation with exceptionally high inflow of water is assigned a factor of 0.05. The joint water reduction factors are given in Table 4.23.

SRF, the stress reduction factor, is a total stress parameter that ranges from 1 to 400, with 1 being most favourable (e.g. rock with unfilled joints) and 400 being most unfavourable (e.g. rock burst). The suggested values of SRF are given in Table 4.24. When the rock mass contains clay, SRF is used to account for the stress relief in excavations and hence loosening of the rock mass (case a in Table 4.24). In competent rock, SRF is a measure of the in situ stress conditions (case b Table 4.24). SRF is also used to account for the squeezing (case c in Table 4.24) and swelling (case d in Table 4.24) loads in plastic-incompetent rocks.

Table 4.23 Joint water reduction J_w

Class	Description	Approx. water pressure (kPa)	J_w
A	Dry excavation or minor inflow (i.e. <5 L/min locally)	<100	1.0
B	Medium inflow or pressure, occasional outwash of joint fillings	100–250	0.66
C	Large inflow or high pressure in competent rock with unfilled joints	250–1000	0.5
D	Large inflow or high pressure, considerable outwash of joint fillings	250–1000	0.33
E	Exceptionally high inflow or water pressure at blasting, decaying with time	>1000	0.2–0.1
F	Exceptionally high inflow or water pressure, continuing without noticeable decay	>1000	0.1–0.05

Source: Barton, N., *Int. J. Rock Mech. Min. Sci.*, 39, 185, 2002.

Source: Barton, N.R., et al., *Rock Mech.*, 6, 189, 1974.

Notes: (1) C to F are crude estimates; increase J_w if drainage measures installed; (2) special problems formed by ice formation are not considered.

Table 4.24 Stress reduction factor SRF

(a) Weakness zones intersecting excavations, which may cause loosening of rock mass when tunnel is excavated

Class	Description	SRF
A	Multiple occurrences of weakness zones containing clay or chemically disintegrated rock, very loose surrounding rock (any depth)	10
B	Single-weakness zone containing clay or chemically disintegrated rock (depth of excavation ≤50 m)	5.0
C	Single-weakness zones containing clay or chemically disintegrated rock (depth of excavation >50 m)	2.5
D	Multiple-shear zones in competent rock (clay-free), loose surrounding rock (any depth)	7.5
E	Single-shear zones in competent rock (clay-free) (depth of excavation ≤50 m)	5.0
F	Single-shear zones in competent rock (clay-free) (depth of excavation >50 m)	2.5
G	Loose, open joints, heavily jointed or sugar cube etc. (any depth)	5.0

Table 4.24 Stress reduction factor SRF (*Continued*)

(b) Competent rock, rock stress problems

Class	*Description*	σ_c/σ_1	σ_θ/σ_c	*SRF*
H	Low stress, near surface, open joints	>200	<0.01	2.5
J	Medium stress, favourable stress condition	200–10	0.01–0.3	1
K	High stress, very tight structure. Usually favourable to stability; may be unfavourable to wall stability	10–5	0.3–0.4	0.5–2
L	Moderate slabbing after >1 h in massive rock	5–3	0.5–0.65	5–50
M	Slabbing and rock burst after a few minutes in massive rock	3–2	0.65–1.0	50–200
N	Heavy rock burst (strain-burst) and dynamic deformations in massive rock	<2	>1	200–400

(c) Squeezing rock: plastic flow of incompetent rock under the influence of high rock pressure

Class	*Description*	σ_θ/σ_c	*SRF*
O	Mild squeezing rock pressure	1–5	5–20
P	Heavy squeezing rock pressure	>5	10–20

(d) Swelling rock: chemical swelling activity depending on presence of water

Class	*Description*	*SRF*
R	Mild swelling rock pressure	5–10
S	Heavy swelling rock pressure	10–15

Source: Barton, N., *Int. J. Rock Mech. Min. Sci.*, 39, 185, 2002.

Source: Barton, N.R., et al., *Rock Mech.*, 6, 189, 1974.

Notes: (1) σ_θ = maximum tangential stress (estimated from elastic theory), σ_c = unconfined compressive strength, σ_1 = major principal stress. (2) Reduce SRF by 25–50% if relevant shear zones only influence but do not intersect the excavation. (3) Barton et al. (1974) have maximum SRF of 20. (4) In strongly anisotropic stress fields (when measured), when σ_1/σ_3 = 5 to 10, reduce σ_c by 25%, and for σ_1/σ_3 > 10, reduce σ_c by 50%.

EXAMPLE 4.6

It is proposed to construct an underground tunnel 500 m below the ground. The drilled cores have an RQD of 85% and the number of joint sets is estimated to be 2. The joints are rough, undulating and unweathered with minor surface staining. The average uniaxial compressive strength of the cores is 190 MPa. The major principal stress acts horizontally and is twice the vertical stress. The unit weight of the rock is approximately 30 kN/m^3. The excavation is relatively dry, with some dampness and negligible inflow. Estimate the Q-value.

Solution

No. of joint sets = 2. Therefore, $J_n = 4$.

Rough and undulating joints → $J_r = 3$

Unaltered joint walls with minor surface staining → $J_a = 1$

Some dampness and negligible inflow → $J_w = 1$

Overburden stress (also, the minor principal stress σ_3) = 30 × 500 kPa = 15 MPa

$\therefore \sigma_1 = 2 \times 15 = 30$ MPa

Uniaxial compressive strength $\sigma_c = 190$ MPa

$\therefore \sigma_c/\sigma_1 = 190/30 = 6.3$

From Table 4.24, SRF = 1.5

Applying Equation 4.10,

$$Q = \left(\frac{85}{4}\right)\left(\frac{3}{1}\right)\left(\frac{1}{1.5}\right) = 42.5$$

On the basis of the Q-value, the rock mass can be classified as shown in Table 4.25. RMR and Q can be approximately related by

$$\text{Bieniawski (1976, 1989): RMR} \approx 9\ln Q + 44 \tag{4.11}$$

$$\text{Barton (1995): RMR} \approx 15\ln Q + 50 \tag{4.12}$$

The Q-value in Equation 4.10 is derived as the product of three quotients. The first one is a measure of the block size. The second is a measure of the joint roughness. The third is a tricky one; it is a stress parameter reflecting the water effects and in situ stresses.

Table 4.25 Rock mass classification for tunnelling work based on Q-system

Q-value	*Class*	*Rock mass quality*
400–1000	A	Exceptionally good
100–400	A	Extremely good
40–100	A	Very good
10–40	B	Good
4–10	C	Fair
1–4	D	Poor
0.1–1.0	E	Very poor
0.01–0.1	F	Extremely poor
0.001–0.01	G	Exceptionally poor

4.7 GEOLOGICAL STRENGTH INDEX

We have looked at the two popular rock mass classification systems, the RMR and Q-systems, which use similar parameters reflecting the intact rock properties and the joint characteristics. They were developed primarily for tunnelling work but are being used for other applications too. The main difference is in the weightings of the relative factors. Uniaxial compressive strength is not a parameter in Q-system; however, it has some influence through the SRF.

The Hoek–Brown failure criterion is quite popular for studying stability of rock mass in underground excavations. In its general form, the failure criterion is expressed as follows:

$$\sigma'_{1f} = \sigma'_{3f} + \sigma_{ci}\left(m_m \frac{\sigma'_{3f}}{\sigma_{ci}} + s\right)^a \tag{4.13}$$

where σ'_{1f} = effective major principal stress at failure, σ'_{3f} = effective minor principal stress at failure and σ_{ci} = uniaxial compressive strength of the intact rock. The constants s and a depend on the rock mass characteristics. The constant s ranges between 0 for poor-quality rock and 1 for intact rock. The constant a ranges between 0.5 for good-quality rock and 0.65 for poor-quality rock. The Hoek–Brown constant m takes separate values of m_i for intact rock and m_m for the rock mass. These are discussed in more detail in Chapter 5. Typical values of m_i for different rock types are given in Table 4.26.

Before 1994, the parameters in the Hoek–Brown criterion were derived from RMR, assuming dry conditions at the excavation (rating increment = 15), with no adjustment for discontinuity orientations with respect to the project (very favourable; rating increment = 0). Noting the fact that relating RMR to Hoek–Brown parameters is not reliable for poor-quality rock masses of low RMR, GSI was introduced in 1994 by Dr. Evert Hoek (Hoek, 1994). It is a number ranging from about 10 for extremely poor-quality rock mass to 100 for extremely strong unjointed rock mass. Around the time of its introduction, GSI was estimated from RMR as follows:

$$\text{GSI} \approx \text{RMR}_{76} \approx \text{RMR}_{89} - 5 \tag{4.14}$$

where RMR_{89} is the value computed according to Bieniawski (1989) as discussed in Section 4.5, and RMR_{76} is the value computed using the older system (Bieniawski, 1976) where the maximum rating increment for groundwater conditions was 10.

GSI is a recent rock mass classification system that was introduced by Hoek (1994) with a heavy reliance on geological observations, and less on numerical values. The two major parameters are (1) surface condition of the discontinuity and (2) interlocking among the rock blocks. The surface

Table 4.26 m_i values for rocks

	Texture			
	Coarse	*Medium*	*Fine*	*Very fine*
Sedimentary	Conglomerates (21 ± 3)	Sandstones, 17 ± 4	Siltstones, 7 ± 2	Claystones, 4 ± 2
	Breccias (19 ± 5)		Greywackes (18 ± 3)	Shales (6 ± 2) Marls (7 ± 2)
	Crystalline limestone (12 ± 3)	Sparitic limestone (10 ± 2)	Micritic limestone (9 ± 2)	Dolomite (9 ± 3)
		Gypsom, 8 ± 2	Anhydrite, 12 ± 2	Chalk, 7 ± 2
Metamorphic	Marble, 9 ± 3	Hornfels (19 ± 4) Metasandstone (19 ± 3)	Quartzite, 20 ± 3	
	Migamatite (29 ± 3)	Amphibiolites, 26 ± 6	Gneiss, 28 ± 5	
		Schists, 12 ± 3	Phyllites (7 ± 3)	Slates, 7 ± 4
Igneous	Granite (32 ± 3)	Diorite (25 ± 5)		
	Granodiorit (29 ± 3)			
	Gabro, 27 ± 3	Dolerite (16 ± 5)		
	Norite, 20 ± 5			
	Porphyries (20 ± 5)		Diabase (15 ± 5)	Peridotite (25 ± 5)
		Rhyolite (25 ± 5) Andesite, 25 ± 5	Dacite (25 ± 3) Basalt (25 ± 5)	Obsidian (19 ± 3)
	Agglomerate (19 ± 3)	Breccia (19 ± 5)	Tuff (13 ± 5)	

Source: Hoek, E., and Brown, E.T., *Int J Rock Mech. Min. Sci. Geomech. Abstr.*, 34, 1165, 1997.

Source: Wyllie, D.C., and Mah, C.W., *Rock Slope Engineering*, 4th edition, Spon Press, London, 2004.

Note: The values in parenthesis are estimates. The others are measured.

condition can vary from 'very good' for fresh unweathered surface to 'very poor' for highly weathered or slickensided surfaces with clay infill. The interlocking blocks can be literally massive at the upper end of the scale to crushed or laminated at the lower end. From these two qualitative parameters, a GSI value is assigned using Figure 4.7.

GEOLOGICAL STRENGTH INDEX FOR JOINTED ROCKS (Hoek and Marinos, 2000) From the lithology, structure and surface conditions of the discontinuities, estimate the average value of GSI. Do not try to be too precise. Quoting a range from 33 to 37 is more realistic than stating that GSI = 35. Note that the table does not apply to structurally controlled failures. Where weak planar structural planes are present in an unfavorable orientation with respect to the excavation face, these will dominate the rock mass behavior. The shear strength of surfaces in rocks that are prone to deterioration as a result of changes in moisture content will be reduced if water is present. When working with rocks in the fair to very poor categories, a shift to the right may be made for wet conditions. Water pressure is dealt with by effective stress analysis. STRUCTURE	SURFACE CONDITIONS VERY GOOD Very rough, fresh unweathered surfaces	GOOD Rough, slightly weathered, iron stained surfaces	FAIR Smooth, moderately weathered and altered surfaces	POOR Slickensided, highly weathered surfaces with compact coatings or fillings or angular fragments	VERY POOR Slickensided, highly weathered surfaces with soft clay coatings or fillings
	DECREASING SURFACE QUALITY ⇒				
DECREASING INTERLOCKING OF ROCK PIECES ⇓					
INTACT OR MASSIVE – intact rock specimens or massive *in situ* rock with few widely spaced discontinuities	90 80			N/A	N/A
BLOCKY – well interlocked undisturbed rock mass consisting of cubical blocks formed by three intersecting discontinuity sets		70 60			
VERY BLOCKY – interlocked, partially disturbed mass with multi-faceted angular blocks formed by 4 or more joint sets			50 40		
BLOCKY/DISTURBED/SEAMY – olded with angular blocks formed by many intersecting discontinuity sets. Persistence of bedding planes or schistosity				30	
DISINTEGRATED – poorly interlocked, heavily broken rock mass with mixture of angular and rounded rock pieces				20	
LAMINATED/SHEARED – lack of blockiness due to close spacing of weak schistosity or shear planes	N/A	N/A			10

Figure 4.7 Geological strength index for jointed rocks.

GSI is one of the parameters used in assessing the strength and deformability of the rock mass using the Hoek–Brown failure criterion. It has been related to m, s and a in Equation 4.13 empirically. The Hoek–Brown parameters for the rock mass and the intact rock are related by

$$m_{\mathrm{m}} = m_{\mathrm{i}} \exp\left(\frac{\mathrm{GSI} - 100}{28}\right) \quad \text{for} \quad \mathrm{GSI} > 25 \tag{4.15}$$

Here, m_{i} is specific to the rock type, and typical values suggested in the literature are given in Table 4.26 (Hoek and Brown, 1997; Wyllie and Mah, 2004). For good-quality rock mass (GSI > 25),

$$a = 0.5 \tag{4.16}$$

$$s = \exp\left(\frac{\mathrm{GSI} - 100}{9}\right) \tag{4.17}$$

Here, the original Hoek–Brown criterion can be used where GSI is estimated from RMR using Equation 4.14. For very poor-quality rock masses, it is difficult to estimate RMR, and hence the modified Hoek–Brown criterion (Hoek et al., 1992) should be used, where GSI has to be estimated from geological observations related to the interlocking of the individual blocks and joint surface conditions as summarised in Table 4.26. For such poor-quality rocks (GSI < 25),

$$a = 0.65 - \frac{\mathrm{GSI}}{200} \tag{4.18}$$

$$s = 0 \tag{4.19}$$

In Chapter 5, you will note that the most recent modification of the Hoek–Brown criterion uses the same expressions for a and s, irrespective of the GSI value. Both approaches give approximately the same values for a and s.

When using the Q-value to derive GSI, as in the case with RMR, it should be assumed that the excavation is dry. A modified Tunnel Quality Index Q' is defined as follows (Hoek et al., 2005):

$$Q' = \left(\frac{\mathrm{RQD}}{J_{\mathrm{n}}}\right)\left(\frac{J_{\mathrm{r}}}{J_{\mathrm{a}}}\right) \tag{4.20}$$

Here, J_{w} and SRF in Equation 4.10 are both taken as 1. Similar to Equation 4.12, GSI can be estimated as

$$\mathrm{GSI} = 9 \ln Q' + 44 \tag{4.21}$$

Table 4.27 Rock mass quality and GSI

GSI	<20	21–40	41–55	56–75	76–95
Rock mass quality	Very poor	Poor	Fair	Good	Very good

Descriptions of rock mass quality, given on the basis of GSI, are shown in Table 4.27.

EXAMPLE 4.7

A granite rock mass has three joint sets, an RQD of 85%, and average joint spacing of 250 mm. Joint surfaces are stepped and rough, unweathered with some stains, and have no separations. The average uniaxial strength of the intact rock cores is 190 kPa, and the excavation area is slightly damp. The excavation is at a depth of 200 m where no unusual in situ stresses are expected. Find the RMR, Q and GSI values. Assume a density of 2.8 t/m^3 and that the vertical in situ stress is the major principal stress.

Assuming dry conditions in the excavation (i.e. maximum rating of 15), compute RMR_{89} and estimate GSI from Equation 4.14

Solution

(a) RMR

UCS = 190 MPa → Rating increment = 12
RQD = 85% → Rating increment = 17
Joint spacing = 250 mm → Rating increment = 10
Joint conditions = Very rough, unweathered, and no separation → Rating increment = 30
Groundwater = Damp → Rating increment = 10
∴ RMR = 12 + 17 + 10 + 30 + 10 = 79

(b) Q

RQD = 85 and $J_n = 9$
Rough and stepped → $J_r = 3$
Unweathered, no separations and some stains → $J_a = 1$
Excavation is damp (i.e. minor inflow) → $J_w = 1$
$\sigma_c = 190$ MPa, $\sigma_v = 200 \times 28/1000 = 5.6$ MPa $\approx \sigma_1$
$\therefore \sigma_c/\sigma_1 = 190/5.6 = 33.9 \rightarrow$ SRF = 1

$$Q = \frac{85}{9} \times \frac{3}{1} \times \frac{1}{1} = 28$$

(c) Rock mass structure = Blocky
Joint surface condition = Very good
GSI = 75 ± 5

(d) Assuming dry conditions in the excavation, $RMR_{89} = 84$.
From Equation 4.11, GSI ≈ 84 – 5 = 79, which matches the value computed in (c).

EXAMPLE 4.8

Check the empirical correlations (Equations 4.11 and 4.12) relating RMR and Q, in light of the RMR and Q values in Example 4.7.

Solution

Substituting $Q = 28$ from Example 4.7 in Equations 4.11 and 4.12,

$$\text{Bieniawski (1989): RMR} \approx 9\ln 28 + 44 = 74$$

$$\text{Barton (1995): RMR} \approx 15\ln 28 + 50 = 100$$

The actual RMR from Example 4.7 is 79. This is in good agreement with the estimate from Equation 4.11 (Bieniawski, 1989).

4.8 SUMMARY

1. The rock mass is weaker and more permeable than the intact rock, mainly due to the discontinuities present.
2. Although all laboratory tests (discussed in Chapter 3) are carried out on the intact rock, it is the strength and deformability of the rock mass that governs the stability.
3. Intact rock strength is only one of the parameters that govern rock mass behaviour.
4. RQD is a reflection of joint spacing or volumetric joint count.
5. The rock mass is classified based on intact rock properties (intact rock strength), joint characteristics (e.g. spacing and roughness) and the boundary conditions (stress field and water).
6. The friction angle ϕ at the joint is derived from two components: the basic friction angle ϕ_b and the joint roughness angle i, that is, $\phi = \phi_b + i$.
7. The basic friction angle ϕ_b is approximately equal to the residual friction angle ϕ_r.
8. RMR and the Tunnelling Quality Index (Q) are the two popular rock mass classification systems. They both rely on similar parameters, with a slight difference in the weightings of these parameters. These were developed for tunnelling, but are used in other applications as well.
9. The first two quotients in Equation 4.10 for computing the Q-value are measures of the block size and joint roughness, respectively. The third quotient is a stress ratio that reflects the effects of water and in situ stresses.
10. GSI is useful in determining the parameters in the Hoek–Brown failure criterion. It is obtained from two qualitative parameters (Table 4.26) describing the interlocking of the rock pieces and the surface quality. It can also be derived indirectly from RMR or Q'.

11. RMR and GSI are numbers that range from 0 to 100. *Q* ranges from 0.001 to 1000+, similar to grain sizes. The larger the value, the better the rock mass characteristics.

Review Exercises

1. State whether the following are true or false.
   a. The larger the RQD, the larger the joint spacing.
   b. The joint roughness is governed more by the large-scale undulations than the small-scale unevenness.
   c. The peak friction angle along a discontinuity can be as high as 70°.
   d. The term aperture applies to both open and filled joints.
2. Carry out a thorough literature review and summarise the empirical correlations relating RMR, *Q* and GSI, including the limitations of the specific correlations.
3. A granite rock formation consists of three sets of discontinuities where the average joint spacing is 320 mm. The RQD of the rock cores obtained from the boreholes is 82%. The joint surfaces are rough, stepped and unweathered, with no separation. The average uniaxial compressive strength of the intact rock cores is 200 MPa. The surface of excavation is found to be damp. Determine the RMR value, disregarding the rating adjustment for discontinuity orientation. What values would you use for cohesion and friction angle in analysis of the rock mass?
   Answer: 79; 390 kPa, 44
4. A tunnel is to be constructed 160 m below ground level, through a highly fractured rock mass where the RQD is 35% and the uniaxial compressive strength of the intact rock cores ranges from 60 to 80 MPa. The joints are separated by 3–4 mm, filled with some clayey silt and are continuous extending to several metres. The joint surfaces are smooth and undulating. Some preliminary measurements show that the groundwater pressure is about 140 m of water and that the overburden pressure can be taken as 160 m of rock. In the absence of any knowledge about the in situ stress ratio, the vertical overburden stress can be taken as the major principal stress. The average unit weight of the rock can be taken as 27 kN/m^3. Estimate the RMR value, without the adjustment for the discontinuity orientation.
   (Hint: *No joint spacing is given. Use RQD in Equation 4.5.*)
   Answer: 34
5. It is proposed to drive a tunnel through a granite rock formation, against the main joint set dipping at 50°. The uniaxial compressive strength of the rock cores tested in the laboratory ranges from 180 to 230 MPa. The RQD of the rock cores is 80%.

The joints are spaced at 500 mm with less than 1 mm separation, and the surfaces are rough and slightly weathered. It is expected that the tunnelling conditions will be wet. Estimate the RMR value with due consideration of the adjustment for discontinuity orientation.In the same location, if the tunnel is driven with the dip, what would be the RMR value?
Answer: 66; 71

6. A sandstone rock mass with RQD of 70% has two joint sets and some random fractures. The joints spaced at 130 mm are generally in contact, with less than 1-mm aperture. The joint surfaces are slightly rough and highly weathered with no clay found on the surface. The uniaxial compressive strength is 95 MPa. The excavation is being carried out at a depth of 110 m below the ground level, and the water table is at a depth of 15 m below the ground level. Estimate RMR, *Q* and GSI. Assume a unit weight of 28 kN/m^3 for the rock mass.
Answer: 52, 4.4, 55 ± 5
7. During an excavation for a tunnel, 250 m below ground level, a highly fractured siltstone rock mass with two major joint sets and many random fractures is encountered. The RQD from the rock cores is 40% and the average UCS is 70 MPa. The joints with average spacing of 75 mm are rather continuous with high persistence with apertures of 3–5 mm, and they are filled with silty clay. The joint surface walls are highly weathered, undulating, and slickensided. There is some water inflow into the tunnel estimated as 15 L/min per 10 m of tunnel length, with some outwash of joint fillings. Estimate RMR, *Q* and GSI. Assume a unit weight of 28 kN/m^3 for the rock mass.
Answer: 40, 1.7, 20 ± 5

REFERENCES

Barton, N. (1973). Review of a new shear strength criterion for rock joints. *Engineering Geology*, Vol. 7, pp. 287–322.

Barton, N.R. (1987). *Predicting the Behaviour of Underground Openings in Rock.* Manuel Rocha Memorial Lecture, Lisbon, NGI Publication 172, Oslo, p. 21.

Barton, N. (2002). Some new Q-value correlations to assist in site characterisation and tunnel design. *International Journal of Rock Mechanics & Mining Sciences*, Vol. 39, No. 2, pp. 185–216.

Barton, N. and Choubey, V. (1977). The shear strength of rock joints in theory and practice. *Rock Mechanics*, Vol. 10, pp. 1–54.

Barton, N.R., Lien, R. and Lunde, J. (1974). Engineering classification of rock masses for the design of tunnel support. *Rock Mechanics*, Vol. 6, No. 4, pp. 189–239.

Bieniawski, Z.T. (1973). Engineering classification of jointed rock masses. *Transactions, South African Institution of Civil Engineers*, Vol. 15, No. 12, pp. 335–344.

Bieniawski, Z.T. (1976). Rock mass classification in rock engineering. *Proceedings of Symposium on Exploration for Rock Engineering*, Ed. Z.T. Bieniawski, A.A. Balkema, Rotterdam, Vol. 1, pp. 97–106.
Bieniawski, Z.T. (1989). *Engineering Rock Mass Classification*, Wiley Interscience, New York, p. 251.
Deere, D.U. and Miller, R.P. (1966). Engineering classification and index properties of intact rock. *Report AFWL-TR-65-116, Air Force Weapon Laboratory (WLDC)*, Kirtland Airforce Base, p. 308.
Franklin, J.A. and Dusseault, M.B. (1989). *Rock Engineering*, McGraw Hill, New York, p. 600.
Hoek, E. (1994). Strength of rock and rock masses. *ISRM News Journal*, Vol. 2, No. 2, pp. 4–16.
Hoek, E. and Brown, E.T. (1997). Practical estimates of rock mass strength. *International Journal of Rock Mechanics and Mining Sciences & Geomechanics Abstracts*, Vol. 34, No. 8, pp. 1165–1186.
Hoek, E., Kaiser, P.K. and Bawden, W.F. (2005). *Support of Underground Excavations in Hard Rock*, A.A. Balkema, Rotterdam.
Hoek, E. and Marinos, P. (2000). Predicting tunnel squeezing. *Tunnels and Tunnelling International*, Part 1, 32/11, pp. 45–51, November 2000; Part 2, 32/12, pp. 33–36, December 2000.
Hoek, E., Wood, D. and Shah, S. (1992). A modified Hoek-Brown criterion for jointed rock masses. *Proceedings of Rock Characterization Symposium*, ISRM, Eurock '92, Ed. J. Hudson, pp. 209–213.
Hudson, J.A. (1989). *Rock Mechanics Principles in Engineering Practice*, Butterworths, London.
Hudson, J.A. and Priest, S.D. (1979). Discontinuities and rock mass geometry. *International Journal of Rock Mechanics and Mining Sciences & Geomechanics Abstracts*, Vol. 16, No. 6, pp. 339–362.
International Society for Rock Mechanics (ISRM) (1978). Suggested methods for quantitative description of discontinuities in rock masses. *International Journal of Rock Mechanics and Mining Sciences & Geomechanics Abstracts*, Vol. 15, No. 6, pp. 319–368.
Laubscher, D. H. (1977). Geomechanics classification of jointed rock masses—mining applications. *Transactions of Institution of Mining and Metallurgy*, Vol. 86, Section A, pp. A1–A8.
Wyllie, D.C. and Mah, C.W. (2004). *Rock Slope Engineering*, 4th edition, Spon Press, London.

Chapter 5

Strength and deformation characteristics of rocks

5.1 INTRODUCTION

Although soils and rocks are both geomaterials, their behaviour under applied loads can be quite different. When it comes to strength and deformation, some of the major differences between rocks and soils are as follows:

- Soils are classic particulate media, and rocks can be seen as a disjointed continuum. There is a significant scale effect in rocks, which is not present in soils. The intact rock, with no structural defects (Figure 5.1a), can be treated as homogeneous and isotropic. On the contrary, the rock mass will often be heterogeneous and anisotropic due to the presence of discontinuities (Figure 5.1b). It can be seen in Figure 5.1b that the stability is better when the loads are applied vertically than horizontally. A highly disjointed or fractured rock (Figure 5.1c) can again be treated as an isotropic material, with a large number of randomly oriented discontinuities. In the case of soils, we generally treat them as homogeneous and isotropic. There are no scale effects in soils; irrespective of the extent considered, the behaviour is the same.
- While the intact rock can show significant tensile strength, the rock mass will have little or no tensile strength due to the presence of discontinuities. We never rely on the tensile strength of soils. However, in good quality rocks with no discontinuities, it is possible to rely on some of its tensile strength.
- Intact rocks have very low porosity with no free water present. Therefore, the permeability is often extremely low. In the rock mass, the discontinuities can contain substantial free water and can lead to high permeabilities. This is known as secondary permeability. The presence of water in the discontinuities can lead to high pore water pressures and hence reduce the effective stresses and shear strength along the discontinuities.

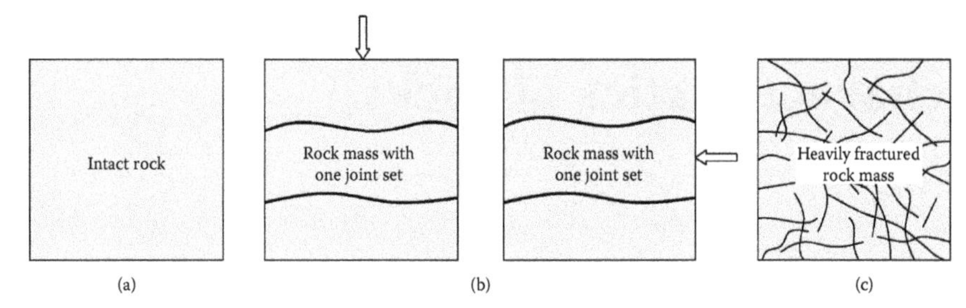

Figure 5.1 Isotropic and anisotropic behaviours: (a) intact rock – isotropic, (b) rock mass with one set of joints – anisotropic and (c) heavily fractured rock mass – isotropic.

- The strength of the intact rock increases with the confining pressure, but not linearly, and it does not follow the Mohr–Coulomb failure criterion very well. The failure stresses are better related by the Hoek–Brown failure criterion, where the failure envelope is parabolic.

5.2 IN SITU STRESSES AND STRENGTH

The overburden stresses within a rock mass are computed the same way as with soils. The unit weight of rocks can be assumed as 27 kN/m^3 in computing the overburden stresses. This value is more than what we normally see in compacted soils or concrete. In situ measurements worldwide, at various depths up to 2500 m, show clearly that the vertical normal stress varies linearly with depth as shown in Figure 5.2a (Hoek and Brown, 1980b). The average in situ vertical overburden stress (σ_v) can be estimated at any depth as

$$\sigma_v \text{ (MPa)} = 0.027\ z(\text{m}) \tag{5.1}$$

where z is the depth in metres.

In normally consolidated and slightly overconsolidated soils, the vertical normal stress is generally the major principal stress and the horizontal stress is the minor principal stress. Here, the *coefficient of earth pressure at rest* K_0, defined as the ratio of horizontal to vertical effective stress, is less than 1. Only in highly overconsolidated soils can K_0 become greater than 1, making the horizontal stresses larger than the vertical stresses. The situation is quite different in rocks, where horizontal stresses are often larger than the vertical stress, especially at depths that are of engineering interest.

In addition to the in situ stresses within the rock mass, stresses are also induced by tectonic activities, erosion and other geological factors. The ratio (K_0) of horizontal normal stress σ_h to vertical normal stress σ_v is generally larger than 1 and can be as high as 3 at shallow depths, where most of the civil engineering works are being carried out. With such a wide variability, horizontal stress should never be estimated. The value of K_0 gets

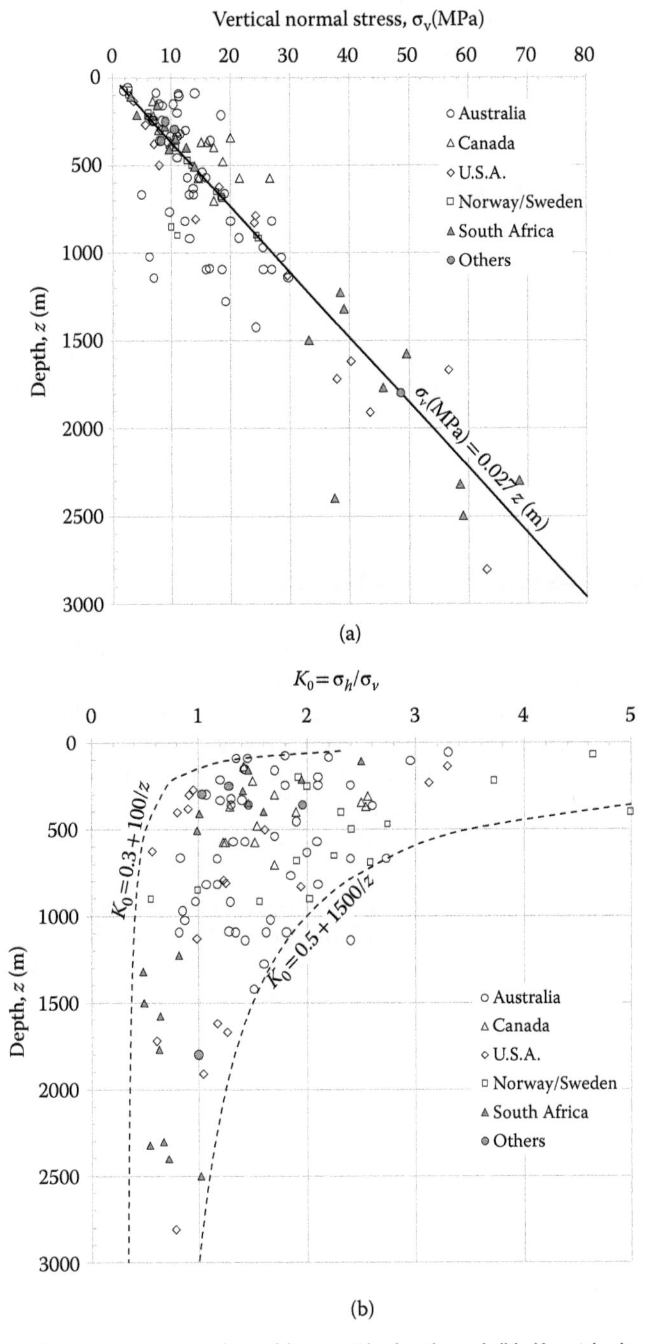

Figure 5.2 In situ measurement data: (a) σ_v with depth and (b) K_0 with depth. (After Hoek and Brown, 1980b.)

smaller with increasing depths. The variations of K_0 values derived from in situ measurements worldwide are plotted against depth in Figure 5.2b. The two dashed lines show the lower and upper bounds for K_0 at any depth. They can be represented by the following equations:

$$\text{Lower bound}: K_0 = 0.3 + \frac{100}{z(\text{m})} \tag{5.2a}$$

$$\text{Upper bound}: K_0 = 0.5 + \frac{1500}{z(\text{m})} \tag{5.2b}$$

Shorey (1994) incorporated the horizontal deformation modulus (E_h) and proposed Equation 5.3 for K_0. The trend and the estimates fit well with those from Hoek and Brown (1980b).

$$K_0 = 0.25 + 7\ E_h(\text{GPa}) \times \left(0.001 + \frac{1}{z(\text{m})}\right) \tag{5.3}$$

5.3 STRESS–STRAIN RELATIONS

The stress–strain relationship of an engineering material is generally specified through a *constitutive relationship* or *constitutive model*. Some of the common constitutive models used to describe the stress–strain behaviour of geomaterials are linear elastic, non-linear elastic, elasto-plastic, elastic-plastic, rigid-plastic, strain hardening, strain softening, Mohr–Coulomb, Cam clay, Drucker–Prager, visco-elastic, visco-plastic and so on. These constitutive models specify how strains are related to stresses.

The simplest analysis of a rock mass is often carried out assuming that it behaves as a *linear isotropic elastic material*, following Hooke's law, which states that strain is proportional to stress. The interrelationships between the six stress components and the six strain components of an isotropic linear elastic material can be written as

$$\varepsilon_x = \frac{1}{E}\left[\sigma_x - \nu\left(\sigma_y + \sigma_z\right)\right] \tag{5.4a}$$

$$\varepsilon_y = \frac{1}{E}\left[\sigma_y - \nu\left(\sigma_x + \sigma_z\right)\right] \tag{5.4b}$$

$$\varepsilon_z = \frac{1}{E}\left[\sigma_z - \nu\left(\sigma_x + \sigma_y\right)\right] \tag{5.4c}$$

$$\gamma_{xy} = \frac{1}{G}\tau_{xy} \tag{5.4d}$$

$$\gamma_{yz} = \frac{1}{G}\tau_{yz} \tag{5.4e}$$

$$\gamma_{zx} = \frac{1}{G}\tau_{zx} \tag{5.4f}$$

where σ = normal stress, τ = shear stress, ε = normal strain and γ = shear strain. x, y and z are the three mutually perpendicular directions in the Cartesian coordinate system. E and G are Young's modulus and the shear modulus (or modulus of rigidity), respectively. ν is Poisson's ratio (see Table 3.3 for typical values), which varies between 0 and 0.5. E and G are related by

$$G = \frac{E}{2(1+\nu)} \tag{3.7}$$

In matrix form, Equation 5.4 can be represented as

$$\begin{Bmatrix} \varepsilon_x \\ \varepsilon_y \\ \varepsilon_z \\ \gamma_{xy} \\ \gamma_{yz} \\ \gamma_{zx} \end{Bmatrix} = \frac{1}{E}\begin{bmatrix} 1 & -\nu & -\nu & 0 & 0 & 0 \\ -\nu & 1 & -\nu & 0 & 0 & 0 \\ -\nu & -\nu & 1 & 0 & 0 & 0 \\ 0 & 0 & 0 & 2(1+\nu) & 0 & 0 \\ 0 & 0 & 0 & 0 & 2(1+\nu) & 0 \\ 0 & 0 & 0 & 0 & 0 & 2(1+\nu) \end{bmatrix}\begin{Bmatrix} \sigma_x \\ \sigma_y \\ \sigma_z \\ \tau_{xy} \\ \tau_{yz} \\ \tau_{zx} \end{Bmatrix} \tag{5.5}$$

or

$$\begin{Bmatrix} \sigma_x \\ \sigma_y \\ \sigma_z \\ \tau_{xy} \\ \tau_{yz} \\ \tau_{zx} \end{Bmatrix} = \frac{E}{(1+\nu)(1-2\nu)}\begin{bmatrix} 1-\nu & \nu & \nu & 0 & 0 & 0 \\ \nu & 1-\nu & \nu & 0 & 0 & 0 \\ \nu & \nu & 1-\nu & 0 & 0 & 0 \\ 0 & 0 & 0 & \frac{(1-2\nu)}{2} & 0 & 0 \\ 0 & 0 & 0 & 0 & \frac{(1-2\nu)}{2} & 0 \\ 0 & 0 & 0 & 0 & 0 & \frac{(1-2\nu)}{2} \end{bmatrix}\begin{Bmatrix} \varepsilon_x \\ \varepsilon_y \\ \varepsilon_z \\ \gamma_{xy} \\ \gamma_{yz} \\ \gamma_{zx} \end{Bmatrix} \tag{5.6}$$

The volumetric strain ε_{vol} is the ratio of volume change to the initial volume and is given by

$$\varepsilon_{vol} = \varepsilon_x + \varepsilon_y + \varepsilon_z \tag{5.7}$$

Substituting the expressions for strains from Equation 5.5,

$$\varepsilon_{vol} = \frac{1-2\nu}{E}\left[\sigma_x + \sigma_y + \sigma_z\right]$$

$$\varepsilon_{vol} = \frac{3(1-2\nu)}{E}\frac{\left[\sigma_x + \sigma_y + \sigma_z\right]}{3} = \frac{1}{K}\frac{\left[\sigma_x + \sigma_y + \sigma_z\right]}{3} \tag{5.8}$$

where K is the bulk modulus given by Equation 3.6 in Chapter 3. In some numerical modelling applications, G and K are used as input parameters rather than E and ν. They are related by

$$E = \frac{9\,KG}{3K+G} \tag{5.9}$$

$$\nu = \frac{3K-2G}{2(3K+G)} \tag{5.10}$$

EXAMPLE 5.1

When the applied normal stresses in the x, y and z directions are principal stresses, express the principal strains in terms of principal stresses, and then the principal stresses in terms of principal strains.

Solution

Substituting $\tau_{xy} = 0$, $\tau_{yz} = 0$ and $\tau_{zx} = 0$ in Equation 5.5,

$$\begin{Bmatrix} \varepsilon_1 \\ \varepsilon_2 \\ \varepsilon_3 \end{Bmatrix} = \frac{1}{E}\begin{bmatrix} 1 & -\nu & -\nu \\ -\nu & 1 & -\nu \\ -\nu & -\nu & 1 \end{bmatrix}\begin{Bmatrix} \sigma_1 \\ \sigma_2 \\ \sigma_3 \end{Bmatrix}$$

Substituting $\gamma_{xy} = 0$, $\gamma_{yz} = 0$ and $\gamma_{zx} = 0$ in Equation 5.6,

$$\begin{Bmatrix} \sigma_1 \\ \sigma_2 \\ \sigma_3 \end{Bmatrix} = \frac{E}{(1+\nu)(1-2\nu)}\begin{bmatrix} 1-\nu & \nu & \nu \\ \nu & 1-\nu & \nu \\ \nu & \nu & 1-\nu \end{bmatrix}\begin{Bmatrix} \varepsilon_1 \\ \varepsilon_2 \\ \varepsilon_3 \end{Bmatrix}$$

5.3.1 Plane strain loading

In geotechnical engineering, when the structure (e.g., retaining wall, embankment and strip footing) is long in one direction, the deformation or strain in this direction can be neglected and the situation can be assumed as a plane strain problem. This is true in rock mechanics too. For a plane strain loading, where the strains are limited to the x-y plane, Equations 5.5 and 5.6 become

$$\begin{Bmatrix} \varepsilon_x \\ \varepsilon_y \\ \gamma_{xy} \end{Bmatrix} = \frac{1}{E} \begin{bmatrix} 1-\nu^2 & -\nu(1+\nu) & 0 \\ -\nu(1+\nu) & 1-\nu^2 & 0 \\ 0 & 0 & 2(1+\nu) \end{bmatrix} \begin{Bmatrix} \sigma_x \\ \sigma_y \\ \tau_{xy} \end{Bmatrix} \tag{5.11}$$

and

$$\begin{Bmatrix} \sigma_x \\ \sigma_y \\ \tau_{xy} \end{Bmatrix} = \frac{E}{(1+\nu)(1-2\nu)} \begin{bmatrix} 1-\nu & \nu & 0 \\ \nu & 1-\nu & 0 \\ 0 & 0 & \frac{(1-2\nu)}{2} \end{bmatrix} \begin{Bmatrix} \varepsilon_x \\ \varepsilon_y \\ \gamma_{xy} \end{Bmatrix} \tag{5.12}$$

Plane strain loading does not mean that there are no normal stresses in the direction perpendicular to the plane. It is the normal *strains* that are zero in that direction. The normal stress in the direction perpendicular to the plane (in the direction of zero normal strain) is given by

$$\sigma_z = \nu\left(\sigma_x + \sigma_y\right) \tag{5.13}$$

In plane strain loading, the non-zero stresses are σ_x, σ_y, σ_z and τ_{xy}. The non-zero strains are ε_x, ε_y and γ_{xy}.

EXAMPLE 5.2

In plane strain loading, when the applied normal stresses in x and y directions are principal stresses, derive the expressions for the major and the minor principal strains.

In a rock mass subjected to plane strain loading, σ_1 = 2 MPa and σ_3 = 1 MPa. Assuming a Young's modulus of 20 GPa and Poisson's ratio of 0.2, determine the principal strains and the normal stress perpendicular to the plane.

Solution

Substituting $\sigma_x = \sigma_1$, $\sigma_y = \sigma_3$ and $\tau_{xy} = 0$ in Equation 5.11, the major and the minor principal strains ε_1 and ε_3 are given by

$$\begin{Bmatrix} \varepsilon_1 \\ \varepsilon_3 \end{Bmatrix} = \frac{1}{E} \begin{bmatrix} 1-\nu^2 & -\nu(1+\nu) \\ -\nu(1+\nu) & 1-\nu^2 \end{bmatrix} \begin{Bmatrix} \sigma_1 \\ \sigma_3 \end{Bmatrix}$$

The principal stress in the direction of zero normal strain is given by

$$\sigma_2 = \nu(\sigma_1 + \sigma_3)$$

This is not necessarily the intermediate principal stress. Depending on the values of ν, σ_1 and σ_3, this can be the minor or intermediate principal stress.

Substituting the values,

$$\varepsilon_1 = \frac{1}{20}\{(1-0.2^2)\times 2 - 0.2\times(1+0.2)\times 1\} = 0.084$$

$$\varepsilon_3 = \frac{1}{20}\{-0.2\times(1+0.2)\times 2 + (1-0.2^2)\times 1\} = 0.024$$

For plane strain loading, $\varepsilon_2 = 0$. For principal planes, the shear strains are zero as well. The normal stress in the direction of zero normal strain is given by

$$\sigma_2 = 0.2\times(2+1) = 0.6 \text{ MPa}$$

5.3.2 Plane stress loading

Plane stress loading is not very common in geotechnical or rock engineering applications. Let us think of a thin plate being loaded along its plane. When the stresses are confined to x-y plane, the stresses and strains are related by

$$\begin{Bmatrix} \varepsilon_x \\ \varepsilon_y \\ \gamma_{xy} \end{Bmatrix} = \frac{1}{E}\begin{bmatrix} 1 & -\nu & 0 \\ -\nu & 1 & 0 \\ 0 & 0 & 2(1+\nu) \end{bmatrix}\begin{Bmatrix} \sigma_x \\ \sigma_y \\ \tau_{xy} \end{Bmatrix} \tag{5.14}$$

or

$$\begin{Bmatrix} \sigma_x \\ \sigma_y \\ \tau_{xy} \end{Bmatrix} = \frac{E}{(1-\nu)^2}\begin{bmatrix} 1 & \nu & 0 \\ \nu & 1 & 0 \\ 0 & 0 & \frac{(1-\nu)}{2} \end{bmatrix}\begin{Bmatrix} \varepsilon_x \\ \varepsilon_y \\ \gamma_{xy} \end{Bmatrix} \tag{5.15}$$

The dimension in the z-direction is very small. Here, the non-zero stresses are σ_x, σ_y and τ_{xy}. There can be strains perpendicular to the x-y plane. The non-zero strains are ε_x, ε_y, ε_z and γ_{xy}. The normal strain in the direction of zero normal stress is given by

$$\varepsilon_z = \frac{\nu}{1-\nu}\left(\varepsilon_x + \varepsilon_y\right) = -\frac{\nu}{E}\left(\sigma_x + \sigma_y\right) \tag{5.16}$$

5.3.3 Axisymmetric loading

Axisymmetric loading is quite common in geotechnical and rock engineering. For example, along the vertical centre line of a uniformly loaded circular footing, the lateral stresses are the same in all directions. If σ_1 and σ_3 are the axial and the radial normal stresses, respectively, they are related to the normal strains in the same directions ε_1 and ε_3 by

$$\begin{Bmatrix} \varepsilon_1 \\ \varepsilon_3 \end{Bmatrix} = \frac{1}{E}\begin{bmatrix} 1 & -2\nu \\ -\nu & 1-\nu \end{bmatrix}\begin{Bmatrix} \sigma_1 \\ \sigma_3 \end{Bmatrix} \tag{5.17}$$

$$\begin{Bmatrix} \sigma_1 \\ \sigma_3 \end{Bmatrix} = \frac{E}{(1+\nu)(1-2\nu)}\begin{bmatrix} 1-\nu & 2\nu \\ \nu & 1 \end{bmatrix}\begin{Bmatrix} \varepsilon_1 \\ \varepsilon_3 \end{Bmatrix} \tag{5.18}$$

5.3.4 Strain–displacement relationships

The strains in the elastic body are caused by displacements. The displacements in the three mutually perpendicular directions u, v and w and strains are related by

$$\begin{Bmatrix} \varepsilon_x \\ \varepsilon_y \\ \varepsilon_z \\ \gamma_{xy} \\ \gamma_{yz} \\ \gamma_{zx} \end{Bmatrix} = \begin{pmatrix} \frac{\partial}{\partial x} & 0 & 0 \\ 0 & \frac{\partial}{\partial y} & 0 \\ 0 & 0 & \frac{\partial}{\partial z} \\ \frac{\partial}{\partial y} & \frac{\partial}{\partial x} & 0 \\ 0 & \frac{\partial}{\partial z} & \frac{\partial}{\partial y} \\ \frac{\partial}{\partial z} & 0 & \frac{\partial}{\partial x} \end{pmatrix}\begin{Bmatrix} u \\ v \\ w \end{Bmatrix} \tag{5.19}$$

5.4 MOHR–COULOMB FAILURE CRITERION

Mohr–Coulomb is the most popular failure criterion that works quite well for geomaterials, especially soils, where the failure generally takes place in shear. The shear strength on the failure plane τ_f is proportional to the normal stress σ on the plane and is expressed as

$$\tau_f = c + \sigma \tan\phi \tag{5.20}$$

where c is the cohesion and ϕ is the friction angle. It can be seen in Equation 5.20 that the shear strength has two separate components: cohesive (c) and frictional ($\sigma \tan \phi$). While the frictional component is proportional to the normal stress, the cohesive component is a constant, which is independent of the normal stress. Let us apply the same Mohr–Coulomb failure criterion to rocks as well.

Uniaxial compression is a very common test that is carried out on soils, rocks and concrete. The uniaxial compressive strength, fondly known as UCS, is denoted as σ_c here. When the specimen fails, $\sigma_1 = \sigma_c$ and $\sigma_3 = 0$. Here, σ_1 and σ_3 are the major and the minor principal stresses, respectively. Uniaxial tensile tests are common on steel specimens but are uncommon for geomaterials or concrete. When testing steel specimens in tension, it is common to use dog-bone-shaped specimens that will prevent slip when the specimen is being pulled axially. This is not possible with rocks. Here, the problem is to hold a specimen without any slippage while the tensile load is applied and increased to failure. Holding the specimen too tight in a chuck would fail the specimen. Further, any misalignment can induce eccentricity and hence a moment, in addition to the axial load. Nevertheless, let us consider a uniaxial tensile strength test, where the magnitude of the tensile strength is σ_t. At failure, $\sigma_1 = 0$ and $\sigma_3 = -\sigma_t$.

The Mohr circles at failure, in a uniaxial tensile strength test and a uniaxial compressive strength test, are shown in Figure 5.3a and b,

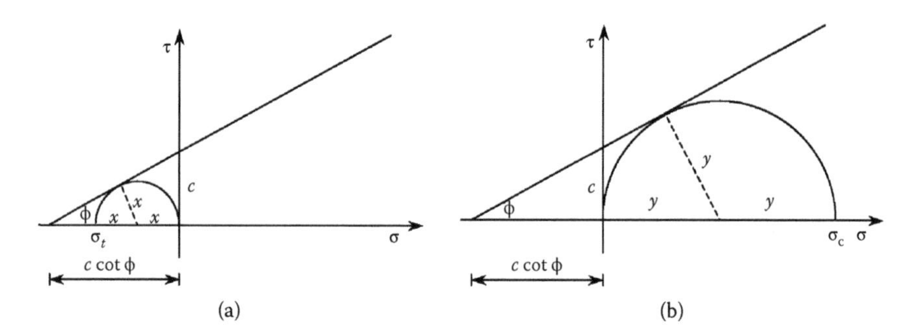

Figure 5.3 Mohr circles at failure: (a) uniaxial tensile strength test and (b) uniaxial compressive strength test.

respectively. Here, it is assumed that the Mohr–Coulomb failure criterion (Equation 5.20) is valid in the tensile region too. By simple trigonometry and algebra, it can be shown that

$$x = \frac{c\cos\phi}{1+\sin\phi}$$

and

$$y = \frac{c\cos\phi}{1-\sin\phi}$$

Noting that the magnitudes of the uniaxial tensile strength σ_t and the uniaxial compressive strength σ_c are given by $2x$ and $2y$, respectively,

$$\sigma_t = \frac{2c\cos\phi}{1+\sin\phi} \tag{5.21}$$

and

$$\sigma_c = \frac{2c\cos\phi}{1-\sin\phi} \tag{5.22}$$

Therefore, the theoretical ratio of σ_c to σ_t of a Mohr–Coulomb material is given by $(1 + \sin\phi)/(1 - \sin\phi)$. For a friction angle of 30–60°, this ratio is in the range of 3–14.

The Brazilian indirect tensile strength test was introduced for rocks and concrete due to the difficulty in carrying out a direct tensile strength test for determining σ_t. In a Brazilian indirect tensile strength test, an intact rock specimen with a thickness to diameter ratio of 0.5 is subjected to a diametrical load P, applied along the entire length of the core, which is increased until failure occurs by splitting (see Chapter 3 for details). Ideally, the disc splits vertically along the diameter into two halves. At failure, the vertical normal stress at the centre of the specimen is compressive and the horizontal normal stress is tensile, as shown in the inset in Figure 5.4. These are also principal stresses. Hondros (1959) showed that the horizontal and the vertical normal stresses at the centre of the core are given by

$$\sigma_{\text{horizontal}}(\text{tensile}) = -\frac{2P}{\pi D t} \tag{5.23}$$

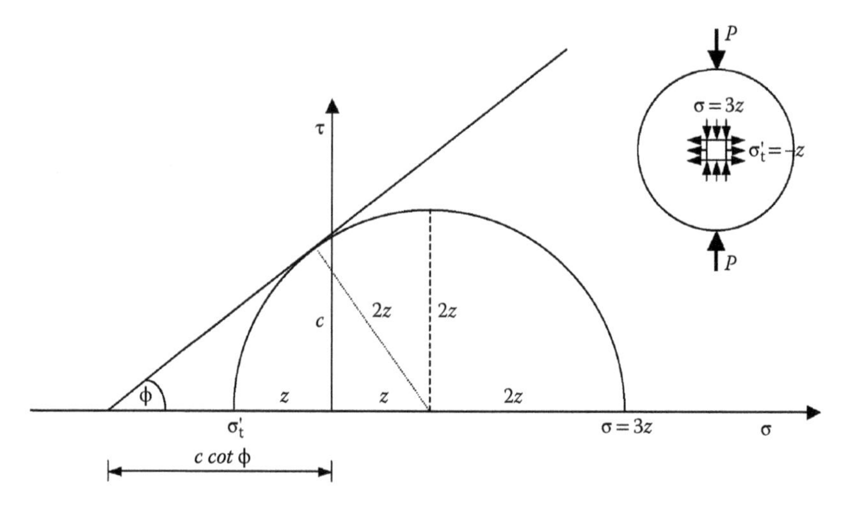

Figure 5.4 Mohr circle for the element at the centre of the specimen during failure in a Brazilian indirect tensile strength test.

$$\sigma_{\text{vertical}}\,(\text{compressive}) = \frac{6P}{\pi D t} \tag{5.24}$$

where P = failure load, D = specimen diameter and t = specimen thickness. The horizontal normal stress at failure, at the centre of the specimen, is known as the indirect tensile strength, denoted here as σ_t'. The million dollar question is how close is it to the uniaxial tensile strength σ_t of the intact rock?

From the Mohr circle shown in Figure 5.4, and by simple trigonometric and algebraic manipulations, it can be shown that

$$z = \frac{c\cos\phi}{2-\sin\phi}$$

and therefore the magnitude of the indirect tensile strength is given by

$$\sigma_t' = \frac{c\cos\phi}{2-\sin\phi} \tag{5.25}$$

which is different from the expression derived for the uniaxial tensile strength σ_t in Equation 5.21. It can be seen here that, theoretically, the magnitude of σ_t' is less than that of σ_t, provided the Mohr–Coulomb criterion is valid in the tensile region as well. The theoretical ratio of σ_c to σ_t' of a Mohr–Coulomb material is given by $2(2 - \sin\phi)/(1 - \sin\phi)$. For a friction angle of 30–60°, this ratio is in the range of 6–17.

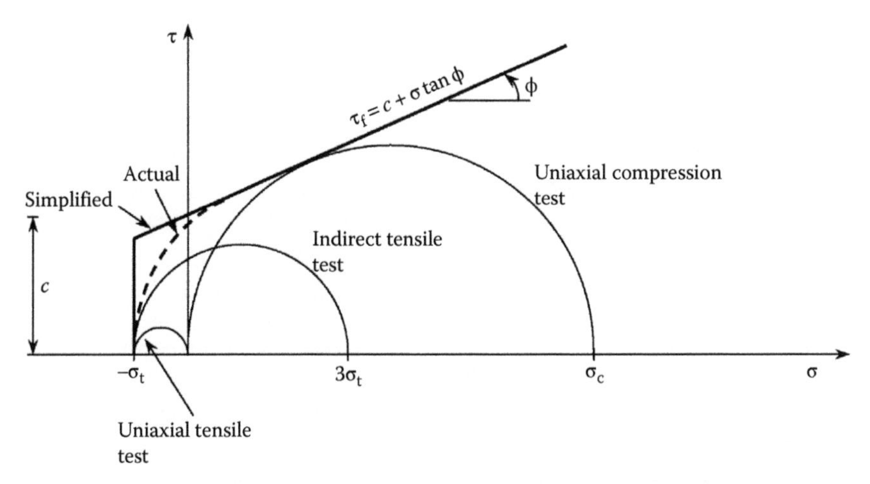

Figure 5.5 Mohr–Coulomb failure criterion with adjustment for tensile normal stresses.

The frictional component in Equation 5.20 is meaningless when the normal stress σ is tensile. Therefore, Equation 5.20 is valid essentially when the normal stress is positive (i.e., compressive). Therefore, in soils, the Mohr–Coulomb failure criterion is used mainly for $\sigma \geq 0$. Unlike soils, rocks can carry some tensile stresses, and therefore the Mohr–Coulomb failure criterion requires some adjustment in the tensile region. There are better failure theories (e.g., Griffith theory) for rocks under tensile stresses.

Figure 5.5 shows a simple extrapolation of the Mohr–Coulomb failure criterion into the tensile region, for situations where the minor principal stress can be negative. Figure 5.5 also shows the three Mohr circles corresponding to the (a) a uniaxial compressive strength test, (b) a Brazilian indirect tensile strength test and (c) a uniaxial tensile strength test. It is assumed that the tensile strength derived from the uniaxial tensile test is the *same* as the one from the Brazilian indirect tensile test. σ_c and σ_t are the uniaxial (or unconfined) compressive and tensile strength values of the rock. The curved dashed line shows the actual envelope in the tensile region, implying that the simplified Mohr–Coulomb extrapolation can overestimate the strength in the tensile region. Therefore, it is prudent to use lower values of c and σ_t when using this simplification (Goodman, 1989). Remember, we are reasonably confident about the Mohr–Coulomb envelope on the right side of the τ-axis; it is the left side that is a worry.

It can be shown from Figure 5.5 that

$$\sigma_c = 2c \tan\left(45 + \frac{\phi}{2}\right) = 2c\frac{\cos\phi}{1 - \sin\phi} = 2c\sqrt{\frac{1 + \sin\phi}{1 - \sin\phi}} \tag{5.26}$$

and, at failure, σ_1 can be related to σ_3 by

$$\sigma_1 = 2c\ \tan\left(45+\frac{\phi}{2}\right)+\sigma_3\tan^2\left(45+\frac{\phi}{2}\right) = \sigma_c+\sigma_3\tan^2\left(45+\frac{\phi}{2}\right) \quad (5.27)$$

EXAMPLE 5.3

Triaxial tests were carried out on 50-mm-diameter limestone cores and the following data were obtained for the principal stresses at failure.

σ_{3f} (MPa)	0	5.0	10.0	20.0	30.0	40.0
σ_{1f} (MPa)	78.0	124.5	145.5	196.0	230.5	262.5

Plot σ_{1f} against σ_{3f} and determine the uniaxial compressive strength σ_c and friction angle ϕ of the limestone.

Solution

From the plot shown in Figure 5.6, σ_c = 95.2 MPa and

$$\tan^2\left(45+\frac{\phi}{2}\right) = 4.4337 \longrightarrow \phi = 39.2°$$

EXAMPLE 5.4

It is proposed to excavate a horseshoe-shaped tunnel at a depth of 1000 m below ground level into a sound unjointed fresh granite with c = 0.5 MPa and ϕ = 40°. The average unit weight of the overburden is 27 kN/m³. Once the tunnel is excavated, it is expected that the lateral normal stress near the tunnel walls will be close to zero. Will the rock fail ('burst') into the excavated tunnel?

What level of prestressing (i.e., σ_3) in the form of struts, rock bolting and so on is required to ensure that the tunnel can just resist the failure?

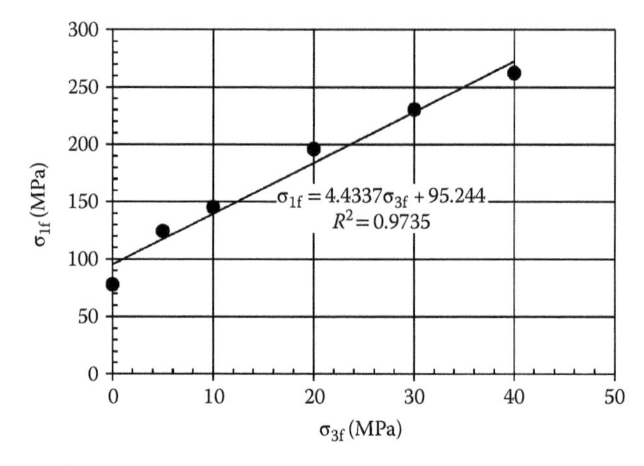

Figure 5.6 Plot of σ_{1f} against σ_{3f}.

Solution

$\sigma_v = \sigma_1 = 100 \times 27$ kPa = 2.7 MPa
$\sigma_h = \sigma_3 \approx 0$ (i.e., more like unconfined compression)
$\phi = 40°$ and $c = 0.5$ MPa

Substituting these values of σ_3, ϕ and c in Equation 5.27, the maximum vertical normal stress the rock can withstand is given by

$$\sigma_{1f} = 2 \times 0.5 \tan\left(45 + \frac{40}{2}\right) + 0 \times \tan^2\left(45 + \frac{40}{2}\right) = 2.14\text{MPa}$$

Since the vertical overburden stress of 2.7 MPa exceeds the available strength of 2.14 MPa, the tunnel wall will fail.

To resist failure, we require some confining pressure σ_3 that would increase the shear strength of the rock. This can be estimated from Equation 5.27 as

$$2.70 = 2 \times 0.5 \tan\left(45 + \frac{40}{2}\right) + \sigma_3 \times \tan^2\left(45 + \frac{40}{2}\right)$$

$$\therefore \sigma_3 = 0.121 \text{ MPa}$$

As in the case of soils, the shear strength of a rock mass can be defined in terms of peak or residual stresses. *Peak shear strength* is the maximum shear stress that can be carried by the element; *residual shear strength*, which is less than the peak shear strength (Figure 5.7a), is the shear stress when the element has undergone significant strain. Using the shear stress values at peak or residual states, Mohr–Coulomb failure envelopes can be developed on the τ–σ plane as shown in Figure 5.7b. Here, the peak and residual friction angles are denoted by ϕ_p and ϕ_r, respectively. In soils and rocks, at

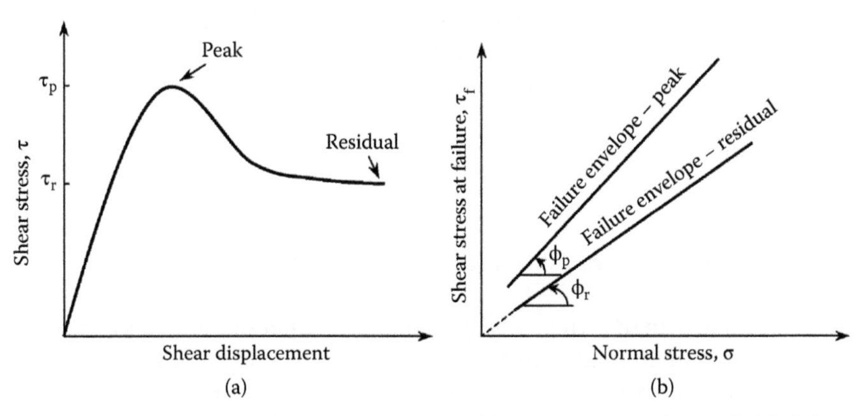

Figure 5.7 Peak and residual shear strengths: (a) stress–strain plot and (b) failure envelopes.

residual states, where the strains are large, the cohesive bonds are broken and there is little or no cohesion contributing towards the shear strength. Therefore, $c_r \approx 0$ and the failure envelope at the residual state passes through the origin in the τ–σ plane. In a rock mass, at large strains, the surface irregularities are further smoothened, and therefore ϕ_r is significantly less than ϕ_p. The difference between the peak friction angle and the residual friction angle can be quite substantial and is approximately equal to the roughness angle *i*, introduced in Equation 4.2. This roughness angle is the result of the surface irregularities or asperities in the rock. The residual friction angle is approximately equal to the basic friction angle of the rock material ϕ_b.

While the intact rock is relatively impervious, the discontinuities present within the rock mass allow easy access to water. With the pore water pressures within the joints, Terzaghi's effective stress theory can be applied to the rock mass. It simply states that the total stress σ is distributed between the rock and the pore water, as effective stress σ' and pore water pressure u, respectively. Therefore,

$$\sigma = \sigma' + u \tag{5.28}$$

Applying the Mohr–Coulomb failure criterion to the *rock mass* in terms of effective stresses, Equation 5.27 becomes

$$\sigma'_{1f} = \sigma_{cm} + \sigma'_{3f}\tan^2\left(45 + \frac{\phi'}{2}\right) = 2c' \tan\left(45 + \frac{\phi'}{2}\right) + \sigma'_{3f}\tan^2\left(45 + \frac{\phi'}{2}\right) \tag{5.29}$$

where σ'_{1f} = effective major principal stress at failure within the rock mass, σ'_{3f} = effective minor principal stress at failure within the rock mass, σ_{cm} = uniaxial compressive strength of the rock mass, c' = effective cohesion of the rock mass and ϕ' = effective friction angle of the rock mass.

We can carry out triaxial tests and determine c' and ϕ' of the intact rock, which is a fairly straightforward exercise. How does one determine c' and ϕ' of the *rock mass*? Ideally, we should test a very large rock mass that includes discontinuities as well. It is a difficult problem. This is discussed further in Section 5.6.

5.5 HOEK–BROWN FAILURE CRITERION

In geotechnical engineering, where the failure within the soil mass occurs in shear, it is common to present the failure criterion in terms of shear and normal stresses on the failure plane. In rock mechanics, however, the common practice is to present the failure criterion in terms of the principal stresses σ_1 and σ_3, having them on *x*- and *y*-axes, respectively.

5.5.1 Intact rock

Noting the deficiencies of the Mohr–Coulomb failure criterion, Hoek and Brown (1980a,b) proposed that the effective major and minor principal stresses within an *intact rock* at failure σ'_{1f} and σ'_{3f}, respectively, can be related by

$$\sigma'_{1f} = \sigma'_{3f} + \sigma_{ci}\left(m_i \frac{\sigma'_{3f}}{\sigma_{ci}} + s\right)^{0.5} \tag{5.30}$$

where s = 1 (for intact rocks only). Here, σ_{ci} is the uniaxial compressive strength of the intact rock, and m_i is the Hoek–Brown parameter for the intact rock, both of which can be determined from a series of triaxial tests. In the past, we used the notation σ_c for uniaxial compressive strength – now we have to separate intact rock and the rock mass and hence give the notation σ_{ci} for the intact rock and σ_{cm} for the rock mass.

For intact rocks, assuming s = 1, Equation 5.30 can be written as

$$\left(\sigma'_{1f} - \sigma'_{3f}\right)^2 = m_i\sigma_{ci}\sigma'_{3f} + \sigma_{ci}^2 \tag{5.31}$$

Plotting the triaxial test data as $\left(\sigma'_{1f} - \sigma'_{3f}\right)^2$ against σ'_{3f}, it is possible to determine m_i and σ_{ci} (see Example 5.5). Alternatively, m_i can be estimated from Table 4.26. It is simply a petrographic constant that is analogous to the friction angle. The strength increases with increasing m_i. The variation of σ'_{1f} against σ'_{3f} as per the Hoek–Brown criterion is shown in Figure 5.8a, where the failure envelope is parabolic. The variation as per the Mohr–Coulomb failure criterion, as deduced from Equation 5.27, is shown in Figure 5.8b, where the failure envelope is a straight line. The intercepts of the σ'_1-axis and the σ'_3-axis are the uniaxial compressive strength σ_{ci} and uniaxial tensile strength σ_{ti}, respectively.

Substituting $\sigma'_{3f} = -\sigma_{ti}$ and $\sigma'_{1f} = 0$ in Equation 5.30,

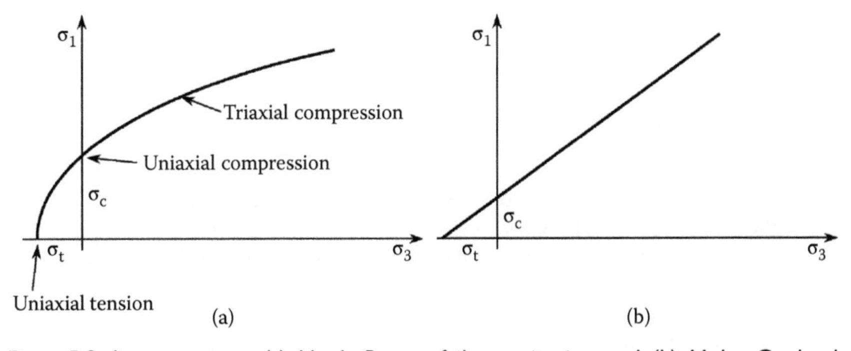

Figure 5.8 A comparison: (a) Hoek–Brown failure criterion and (b) Mohr–Coulomb failure criterion.

$$0 = -\sigma_{ti} + \sigma_{ci}\left(-m_i \frac{\sigma_{ti}}{\sigma_{ci}} + s\right)^{0.5}$$

$$\left(\frac{\sigma_{ti}}{\sigma_{ci}}\right)^2 + m_i\left(\frac{\sigma_{ti}}{\sigma_{ci}}\right) - s = 0$$

Therefore,

$$\left(\frac{\sigma_{ti}}{\sigma_{ci}}\right) = -\frac{\sqrt{m_i^2 + 4s} - m_i}{2} \tag{5.32}$$

Equation 5.32 shows that the ratio of compressive strength to tensile strength of an intact rock depends only on m_i. This ratio σ_{ci}/σ_{ti} increases with m_i. For the range of m_i = 5–35, σ_{ci}/σ_{ti} lies within 5 and 35. As a first approximation, σ_{ci}/σ_{ti} can be taken as m_i.

EXAMPLE 5.5

Triaxial tests were carried out on 50-mm-diameter limestone cores and the following data were obtained for the principal stresses at failure (same data as in Example 5.3).

σ_{3f} (MPa)	0	5.0	10.0	20.0	30.0	40.0
σ_{1f} (MPa)	78.0	124.5	145.5	196.0	230.5	262.5

Neglect the pore water pressures. Plot $(\sigma_{1f} - \sigma_{3f})^2$ versus σ_{3f} and determine m_i and σ_{ci} for the limestone.

Using the preceding values of m_i and σ_{ci}, plot the theoretical failure envelope in σ_{3f} versus σ_{1f} space.

Show the test data along with the theoretical failure envelope and see how well they match.

Solution

The plot of $(\sigma_{1f} - \sigma_{3f})^2$ versus σ_{3f} is shown in Figure 5.9. From the line of best fit,

$\sigma_{ci}^2 = 7835 \longrightarrow \therefore \sigma_{ci} = 88.5$ MPa
The gradient $m_i\sigma_{ci} = 1070.4 \longrightarrow m_i = 12.1$

Substituting $m_i = 12.1$, $\sigma_{ci} = 88.5$ MPa and $s = 1$ in Equation 5.30, the theoretical Hoek–Brown failure envelope can be derived.

The theoretical envelope derived here is shown along with the experimental data in Figure 5.10.

EXAMPLE 5.6

In Example 5.5, estimate the tensile strength of the rock.

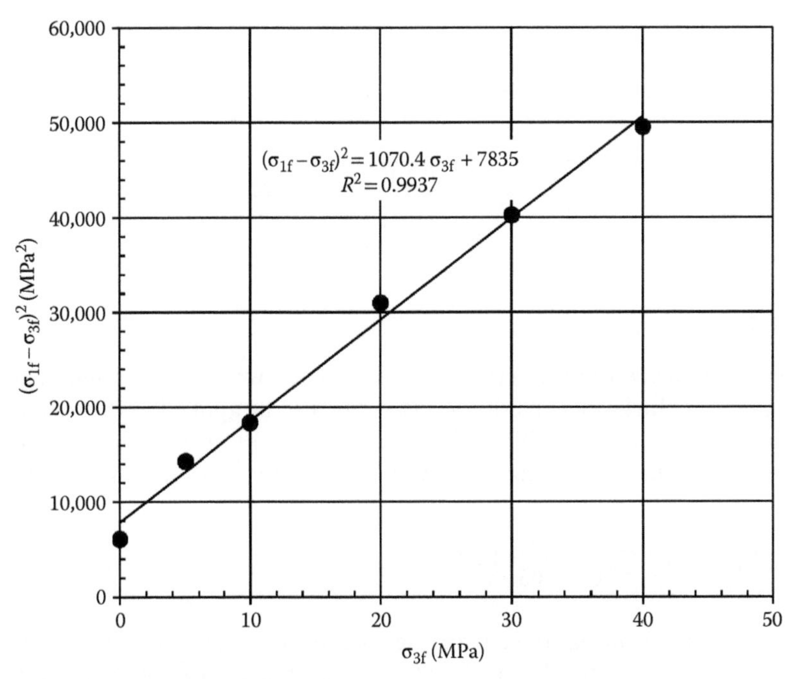

Figure 5.9 Determination of σ_{ci} and m_i.

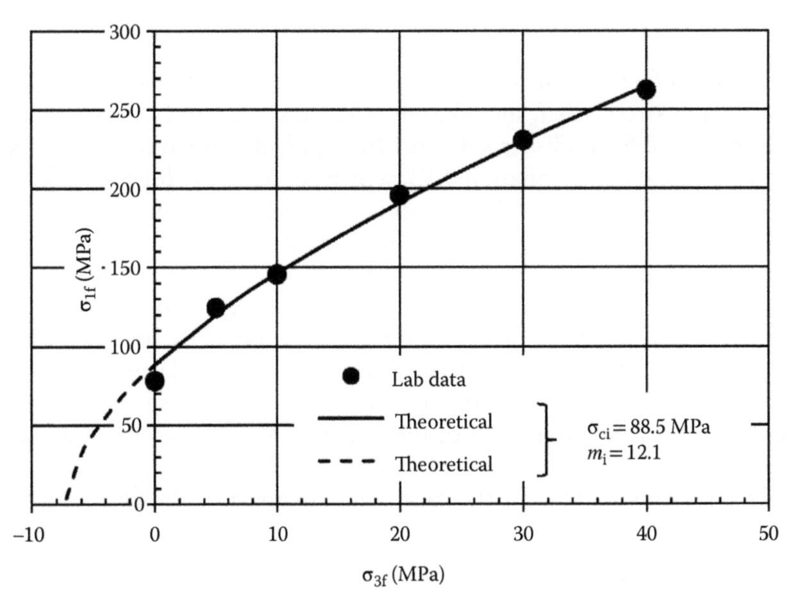

Figure 5.10 Theoretical Hoek–Brown failure envelope.

Solution

From the triaxial test data, it was determined in Example 5.5 that m_i = 12.1 and σ_{ci} = 88.5 MPa. Assuming s = 1 and substituting in Equation 5.32,

$$\left(\frac{\sigma_{ti}}{\sigma_{ci}}\right) = -\frac{\sqrt{m_i^2 + 4s} - m_i}{2} = -\frac{\sqrt{12.1^2 + 4 \times 1} - 12.1}{2} = -0.082$$

$$\therefore \sigma_{ti} = -7.3 \text{ MPa}$$

5.5.2 Rock mass

The Hoek–Brown failure criterion has evolved over the years into a more *generalised* Hoek–Brown criterion that can also be applied to the rock mass as well as intact rocks. This was discussed briefly in Chapter 4. For the jointed *rock mass*, Equation 5.30 was modified to (Hoek and Brown, 1997)

$$\sigma'_{1f} = \sigma'_{3f} + \sigma_{ci}\left(m_m \frac{\sigma'_{3f}}{\sigma'_{ci}} + s\right)^a \tag{4.13}$$

where m_m is the m-parameter for the rock mass (sometimes denoted as m_b in the literature with subscript b referring to *broken rock*), which is derived from the value for the intact rock m_i as (Hoek et al., 2002)

$$m_m = m_i \; exp\left(\frac{\text{GSI} - 100}{28 - 14D}\right) \tag{5.33}$$

where D is a factor to account for the disturbance in the rock mass due to blasting and stress relief, introduced by Hoek et al. (2002). It varies in the range of 0–1; 0 for undisturbed and 1 for highly disturbed rock mass. Note that we still use σ_{ci} in Equation 4.13, which is essentially for the rock mass.

The constant m_m can take a positive value in the range of 0.001–25, with highly disturbed poor quality rock masses falling in the lower end and the hard and almost intact rocks at the upper end. It can be seen from Equation 5.33 that m_m is less than m_i, which is expected intuitively. Yes, the rock mass is weaker than the intact rock. Typically, m_i varies in the range of 2–35. The difference between the two becomes larger with poorer quality rock mass with low GSI. The uniaxial compressive strength of the rock mass σ_{cm} is less than that of the intact rock σ_{ci} due to the presence of discontinuities.

The constants s and a for the rock mass are given by (Hoek et al., 2002)

$$s = \exp\left(\frac{\text{GSI} - 100}{9 - 3D}\right) \tag{5.34}$$

and

$$a = \frac{1}{2} + \frac{1}{6}\left(e^{-GSI/15} - e^{-20/3}\right) \quad (5.35)$$

Generally, s varies in the range of 0–1, mostly in the lower end of the range, with 0 for very poor quality rock and 1 for intact rock. It is a petrographic constant that is similar to cohesion in the Mohr–Coulomb failure criterion. The constant a varies between 0.50 for good quality rock (or intact rock) and 0.65 for poor quality rock. D is a factor to account for the disturbance within the rock mass due to blasting, stress relief and so on. It varies in the range of 0–1; 0 for undisturbed and 1 for highly disturbed rock mass.

The Hoek–Brown failure criterion is developed assuming isotropic behaviour of the intact rock and rock mass. Therefore, it works well for intact rock specimens as well as closely spaced heavily jointed rock masses where isotropy can be assumed. In situations where the structure being analysed and the block sizes are of the same order in size, or in situations with specific weak discontinuities, the Hoek–Brown failure criterion should not be applied.

Substituting $\sigma'_{3f} = 0$ and $\sigma'_{1f} = \sigma_{cm}$ in Equation 4.13, the uniaxial compressive strength of the rock mass can be calculated as

$$\sigma_{cm} = \sigma_{ci} s^a \quad (5.36)$$

where σ_{ci} is the UCS of the intact rock. Marinos and Hoek (2001) proposed an empirical equation for σ_{cm} in terms of m_i, σ_{ci} and GSI as

$$\sigma_{cm} = \sigma_{ci} \times 0.0034\, m_i^{0.8} \times \{1.029 + 0.025e^{-0.1 m_i}\}^{GSI} \quad (5.37)$$

The ratio σ_{cm}/σ_{ci} approaches unity when GSI increases to 100. There are empirical equations reported in the literature that relate σ_{cm}/σ_{ci} to RMR or Q.

Assuming $a = 0.5$, Equation 5.32 can be extended to the rock mass, where the uniaxial tensile strength can be expressed as (Hoek and Brown, 1997)

$$\sigma_{tm} = -\sigma_{ci} \frac{\sqrt{m_m^2 + 4s} - m_m}{2} \quad (5.38)$$

Hoek (1983) noted that for brittle materials, the uniaxial tensile strength is the same as the biaxial tensile strength. Therefore, substituting $\sigma'_{3f} = \sigma'_{1f} = \sigma_{tm}$ in Equation 4.13, the tensile strength of the rock mass is given by (Hoek et al., 2002)

Table 5.1 Some typical values of Hoek–Brown parameters from case histories

	Intact Rock		*Rock Mass*				
Description	σ_{ci} *(MPa)*	m_i	*GSI*	m_m	s	a	E_m *(GPa)*
Massive (almost intact) but weak cemented breccias – similar to weak concrete	51	16.3	75	6.68	0.062	0.501	15.0
Massive strong rock gneiss, with very few joints	110	28	75	11.46	0.062	0.501	45.0
Average quality rock mass: jointed quartz mica schist	30	15	65	4.3	0.02	0.5	10.0
Poor quality rock mass at shallow depth: Athenian schist	5–10	9.6	20	0.55	0.0001	0.544	0.60
Poor quality rock mass under high stress: 25-km-long water supply tunnel 1200 m below surface; graphitic phyllite, squeezing ground	50	10	25	0.48	0.0002	0.53	1.0

Source: Hoek, E., *Practical Rock Engineering*, http://www.rocscience.com/hoek/corner/practical_rock_engineering.pdf, 2007.

$$\sigma_{\rm tm} = -\frac{s\sigma_{\rm ci}}{m_{\rm m}} \tag{5.39}$$

Some typical values of the Hoek–Brown parameters of the intact rock and the rock mass, the GSI of the rock mass and the deformation modulus of the rock mass from a few larger projects worldwide, as documented by Hoek (2007), are summarised in Table 5.1.

EXAMPLE 5.7

For the massive strong rock gneiss with very few joints in Table 5.1, with GSI = 75, m_i = 28 and σ_{ci} = 110 MPa, estimate the rock mass parameters m_m, s and a. How do they compare with the values given in Table 5.1?

Estimate the uniaxial compressive strength of the rock mass σ_{cm}.

Solution

Assuming $D = 0$,

From Equation 5.33 ⟶ $m_m = 11.46$.
From Equation 5.34 ⟶ $s = 0.062$.
From Equation 5.35 ⟶ $a = 0.501$.

The values match those in Table 5.1 very well.
Substituting in Equation 5.36, $\sigma_{cm} = \sigma_{ci} s^a = 110 \times 0.062^{0.501} = 27.3$ MPa.

5.6 MOHR–COULOMB c' AND ϕ' FOR ROCK MASS FROM THE HOEK–BROWN PARAMETERS

We have seen in Section 5.5 that deriving the Hoek–Brown parameters for the rock mass from those of an intact rock is a straightforward exercise. For the intact rock, the parameters are m_i and σ_{ci} ($s = 1$ and $a = 0.5$). For the rock mass, the parameters are m_m, σ_{cm}, s and a. These two sets of parameters are related by GSI and D that reflect the quality of the rock mass and the degree of disturbance it has undergone during excavation, blasting and so on.

The Mohr–Coulomb failure criterion is quite popular among geotechnical engineers, and there is a tendency to apply this to rocks too. The main difficulty here is to derive the shear strength parameters c' and ϕ' in terms of effective stresses for the *rock mass*. It is not practical to test a representative rock mass in a triaxial cell. It can only be carried out through a simulation exercise.

Hoek and Brown (1997) simulated a series of triaxial test data for the rock masses of different GSI, m_i and σ_{ci} values, in the confining pressure σ'_{3f} range of $0–\sigma_{ci}/2$. Mohr–Coulomb envelopes were drawn with these simulated data from which c' and ϕ' for the rock masses were determined. The values of c' and ϕ' thus determined are presented graphically in Figures 5.11 and 5.12. It should be noted that the synthetic data were generated to follow a parabolic failure envelope in $\sigma'_1–\sigma'_3$ space. The linear Mohr–Coulomb envelope fitted to these data will vary depending on the stress range covered ($\sigma_{tm} < \sigma'_{3f} < \sigma'_{3,max}$). Therefore, the Mohr–Coulomb parameters c' and ϕ' will vary depending on the range of values selected for σ'_{3f}. A simple simulation exercise is shown through Examples 5.8 and 5.9.

Hoek et al. (2002) reported that the curve fitting exercise gives the following expressions for determining ϕ' and c'.

$$\sin\phi' = \frac{6am_m(s+m_m\sigma'_{3n})^{a-1}}{2(1+a)(2+a)+6am_m(s+m_m\sigma'_{3n})^{a-1}} \tag{5.40}$$

$$\frac{c'}{\sigma_{ci}} = \frac{\left[(1+2a)s+(1-a)m_m\sigma'_{3n}\right](s+m_m\sigma'_{3n})^{a-1}}{(1+a)(2+a)\sqrt{1+\left(6am_m(s+m_m\sigma'_{3n})^{a-1}\right)/\left((1+a)(2+a)\right)}} \tag{5.41}$$

where $\sigma'_{3n} = \sigma'_{3,max}/\sigma_{ci}$. They also suggested that $\sigma'_{3,max}$, the upper limit of σ'_{3f}, should be selected depending on the project and stress levels. As a general guide, $\sigma'_{3,max}$ can be estimated from the following equation for tunnels and underground excavations (Hoek et al., 2002):

$$\frac{\sigma'_{3,max}}{\sigma'_{cm}} = 0.47\left(\frac{\sigma'_{cm}}{\gamma H}\right)^{-0.94} \tag{5.42}$$

where H is the depth below the surface and γ is the unit weight of the rock mass. σ'_{cm} is what Hoek and Brown (1997) refer to as the *global rock mass strength*, determined from the Mohr–Coulomb failure envelope fitted to the simulated data. It is a better representation of the average uniaxial compressive strength of the rock mass. This is simply the uniaxial compressive strength determined from the Mohr–Coulomb failure criterion, which is generally larger than the rock mass strength σ_{cm} (Equation 5.36) determined from the Hoek–Brown criterion. For slopes, $\sigma'_{3,max}$ can be estimated from (Hoek et al., 2002)

$$\frac{\sigma'_{3,max}}{\sigma'_{cm}} = 0.72\left(\frac{\sigma'_{cm}}{\gamma H}\right)^{-0.91} \tag{5.43}$$

where H = height of the slope. Equation 5.43 was developed assuming two-dimensional failure surfaces in the form of circular arcs and Bishop's method of slices.

From the Mohr–Coulomb envelope,

$$\sigma'_{cm} = \frac{2c'\cos\phi'}{1-\sin\phi'} \tag{5.44}$$

where c' and ϕ' are the shear strength parameters for the rock mass in terms of effective stresses. In the normal stress range of $\sigma_t < \sigma'_{3f} < 0.25\sigma_{ci}$ (Hoek et al., 2002),

$$\sigma'_{cm} = \sigma_{ci}\frac{\left[m_b + 4s - a(m_b - 8s)\right](0.25m_b + s)^{a-1}}{2(1+a)(2+a)} \tag{5.45}$$

EXAMPLE 5.8

Let us carry out a simple simulation exercise. Generate a set of triaxial data for the *rock mass* in Example 5.7, by determining the values of σ'_{1f} for σ'_{3f} = 0, 10, 20, 40 and 60 MPa.

Solution

For the rock mass,

$$\sigma'_{1f} = \sigma'_{3f} + \sigma_{ci}\left(m_b \frac{\sigma'_{3f}}{\sigma'_{ci}} + s\right)^a$$

$$\sigma'_{1f} = \sigma'_{3f} + 110\left(11.46\frac{\sigma'_{3f}}{110} + 0.062\right)^{0.501}$$

Substituting for σ'_{3f} in the preceding equation, the following values are obtained for σ'_{1f}.

σ'_{3f} (MPa)	0	10	20	40	60
σ'_{1f} (MPa)	27.3	125.6	181.3	266.5	336.9

EXAMPLE 5.9

Use the synthetic triaxial data for the rock mass from Example 5.8 and draw the Mohr–Coulomb envelope in the σ'_{1f}–σ'_{3f} plane. Determine c' and ϕ' and check whether the values match those estimated from Figures 5.11 and 5.12.

Solution

The experimental data and the Mohr–Coulomb envelope are shown in Figure 5.13.

From Equation 5.29, $\tan^2\left(45 + \frac{\phi'}{2}\right) = 4.9045 \longrightarrow \phi' = 41.4°$ and

$$2c' \tan\left(45 + \frac{\phi'}{2}\right) = 60.003 \longrightarrow c' = 13.5 \text{ MPa}$$

For GSI = 75 and m_i = 28 (see Example 5.7). From Figure 5.11, $c'/\sigma_{ci} = 0.086 \longrightarrow c' = 9.5$ MPa; from Figure 5.12, $\phi' = 47°$.·

There are some difference between the computed values and those estimated from Figures 5.11 and 5.12.

Note that fitting a straight-line envelope to satisfy the Mohr–Coulomb criterion gives a global rock mass strength σ'_{cm} of 60 MPa, which is greater than the σ_{cm} estimated as 27.3 MPa in Example 5.7.

5.7 DEFORMATION MODULUS

The deformation modulus of the rock mass is a very important parameter in computing the strains or deformations. The Young's modulus of the intact rock (E_i) is generally derived from uniaxial compression tests on the intact cores. In the absence of laboratory measurements, E_i can be estimated from an assumed value of σ_c and modulus ratio (E/σ_c), which varies in the range

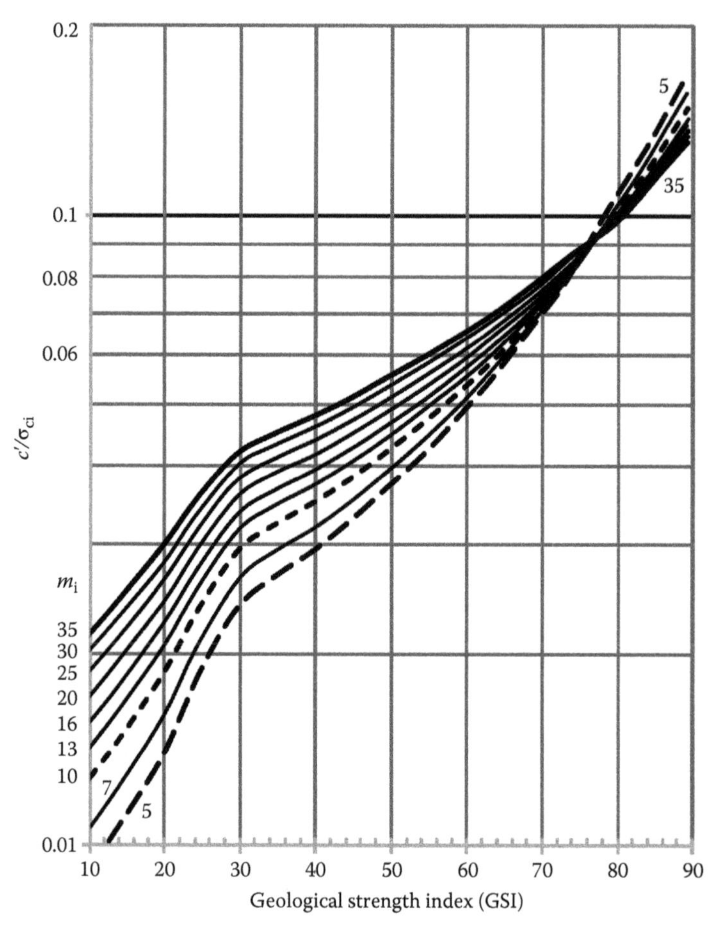

Figure 5.11 c'/σ_{ci}–m_i–GSI relationship for the rock mass. (After Hoek, E. and E.T. Brown, *Int. J. Rock Mech. Min. Sci.*, 34, 1165–1186, 1997.)

of 150–1000. Typical values of modulus ratio and σ_c are given in Tables 3.5 and 3.6, respectively. The rock mass modulus can be determined from empirical correlations discussed in this section.

The Young's modulus (E_i) of an intact rock is generally 150–1000 times the uniaxial compressive strength. It is generally measured at low stress levels where the rock behaves elastically. The stiffness (i.e., Young's modulus) is fairly consistent for a rock type even though there can be some scatter in the strength data. It can vary from less than 1 GPa to more than 100 GPa (see Figure 3.8 and Table 3.6).

The deformation modulus can be estimated from the tunnel quality index Q as (Grimstad and Barton, 1993)

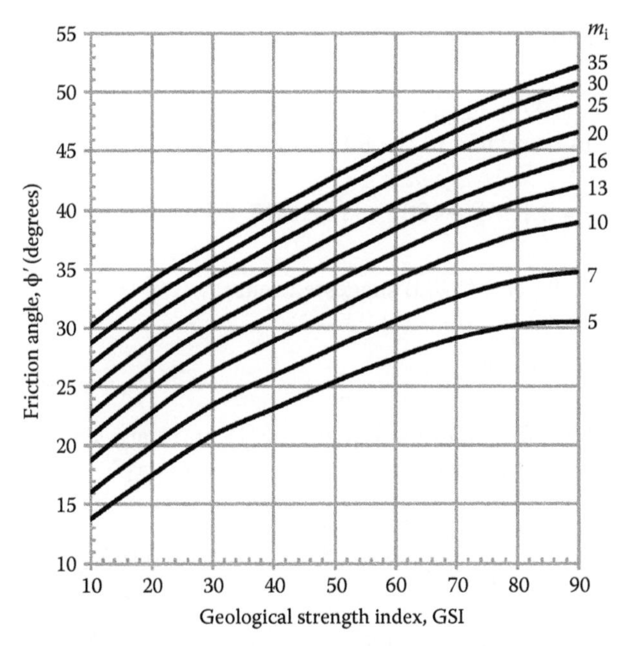

Figure 5.12 φ′–m_i–GSI relationship for the rock mass. (After Hoek, E. and E.T. Brown, *Int. J. Rock Mech. Min. Sci.*, 34, 1165–1186, 1997.)

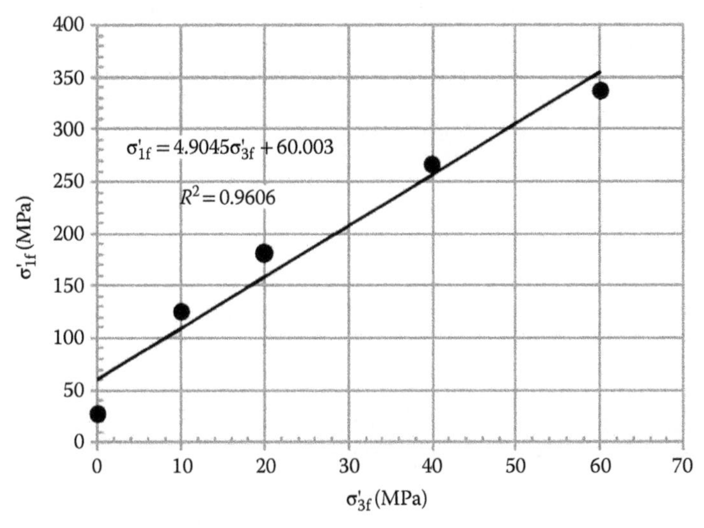

Figure 5.13 Mohr–Coulomb envelope in the σ'_{1f}–σ'_{3f} plane.

$$E_m = 25\log Q \qquad \text{for } Q > 1 \tag{5.46}$$

Bieniawski (1978) suggested that the in situ deformation modulus of a rock mass can be related to the RMR by

$$E_m\,(\text{GPa}) = 2\ \text{RMR} - 100 \qquad \text{for RMR} > 55 \tag{5.47}$$

Serafim and Pereira (1983) suggested that

$$E_m\,(\text{GPa}) = 10^{\frac{\text{RMR}-10}{40}} \tag{5.48}$$

Hoek et al. (2002) modified Equation 5.48 and suggested that the deformation modulus of the rock mass can be expressed as

$$E_m\,(\text{GPa}) = \left(1 - \frac{D}{2}\right)\sqrt{\frac{\sigma_{ci}}{100}} \times 10^{\frac{\text{GSI}-10}{40}} \qquad \text{for } \sigma_{ci} < 100 \text{ MPa} \tag{5.49a}$$

$$E_m\,(\text{GPa}) = \left(1 - \frac{D}{2}\right) \times 10^{\frac{\text{GSI}-10}{40}} \qquad \text{for } \sigma_{ci} > 100 \text{ MPa} \tag{5.49b}$$

Hoek and Diederichs (2006) reviewed several empirical relationships that are used to estimate the deformation modulus of the rock mass. Based on a large number of in situ measurements from China and Taiwan, they proposed the following two equations:

$$E_m\ (\text{GPa}) = 100\left(\frac{1 - D/2}{1 + e^{(75+25D-\text{GSI})/11}}\right) \tag{5.50}$$

$$E_m = E_i\left(0.02 + \frac{1 - D/2}{1 + e^{(60+15D-\text{GSI})/11}}\right) \tag{5.51}$$

From Equation 5.50, it is evident that the ratio E_m/E_i approaches unity with GSI increasing towards 100.

5.8 STRENGTH OF ROCK MASS WITH A SINGLE PLANE OF WEAKNESS

Let us consider the simple situation shown in Figure 5.14a, where the rock mass consists of a single joint, inclined at angle β to the major principal plane. The major and the minor principal stresses are σ_1 and σ_3, respectively.

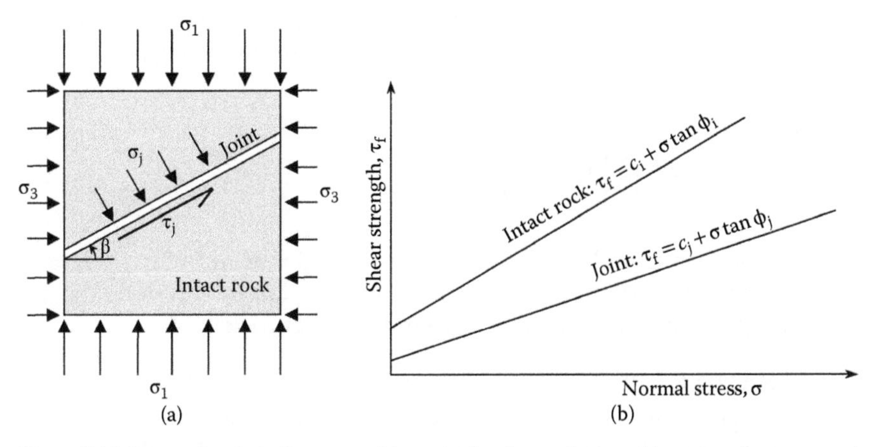

Figure 5.14 Strength of rock mass with a single discontinuity: (a) state of stress and (b) failure envelopes.

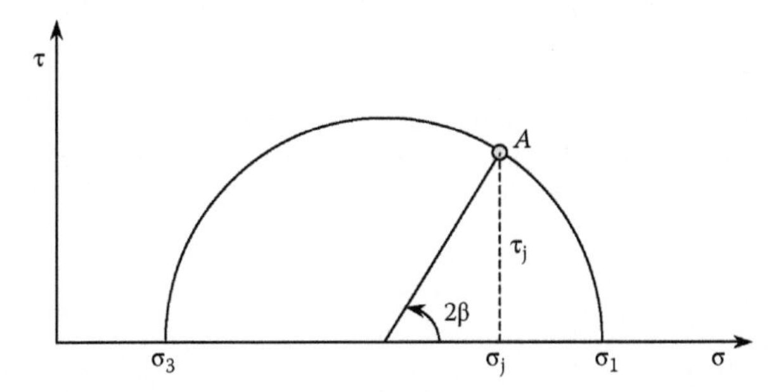

Figure 5.15 Mohr circle representing the state of stress shown in Figure 5.14a.

The discontinuity planes are often weaker than the intact rock with a lower cohesion and friction angle (Figure 5.14b).

The Mohr circle representing the state of stress at the intact rock (Figure 5.14a) is shown in Figure 5.15. The shear and normal stresses at the discontinuity plane (i.e., joint) are represented by point A and are given by

$$\sigma_j = \sigma_3 + \left(\frac{\sigma_1 - \sigma_3}{2}\right)(1 + \cos 2\beta) \tag{5.52}$$

$$\tau_j = \left(\frac{\sigma_1 - \sigma_3}{2}\right)\sin 2\beta \tag{5.53}$$

For failure to take place along the joint, these two values of σ_j and τ_j should satisfy the equation representing the Mohr–Coulomb failure envelope given by

$$\tau_j = c_j + \sigma_j \tan_j \phi_j \tag{5.54}$$

Substituting the expressions for σ_j (Equation 5.52) and τ_j (Equation 5.53) in Equation 5.54, it can be shown that when slip occurs along the joint

$$(\sigma_1 - \sigma_3) = \frac{2(c_j + \sigma_3 \tan\phi_j)}{\sin 2\beta \left(1 - \dfrac{\tan\phi_j}{\tan\beta}\right)} \tag{5.55}$$

It can be seen from Equation 5.55 that when $\beta = \phi_j$ or $90°$, $\sigma_1 - \sigma_3 = \infty$. Under such circumstances, the rock mass will not fail by slip along the discontinuity; failure can only take place in the intact rock.

EXAMPLE 5.10

A large extent of rock mass has a single plane of discontinuity, where the aperture is filled. The shear strength parameters for this fill material are $c = 4.0$ MPa and $\phi = 34°$. For $\sigma_3 = 3$ MPa, find the values of σ_1 for different values of β and plot the variation of σ_1 against β.

Solution

The variation of σ_1 against β is shown in Figure 5.16. For $\beta < \phi$, slip is not possible. When the discontinuity is oriented at an angle less than ϕ (= 34° in this case) to the principal plane, failure can only take place in the intact rock.

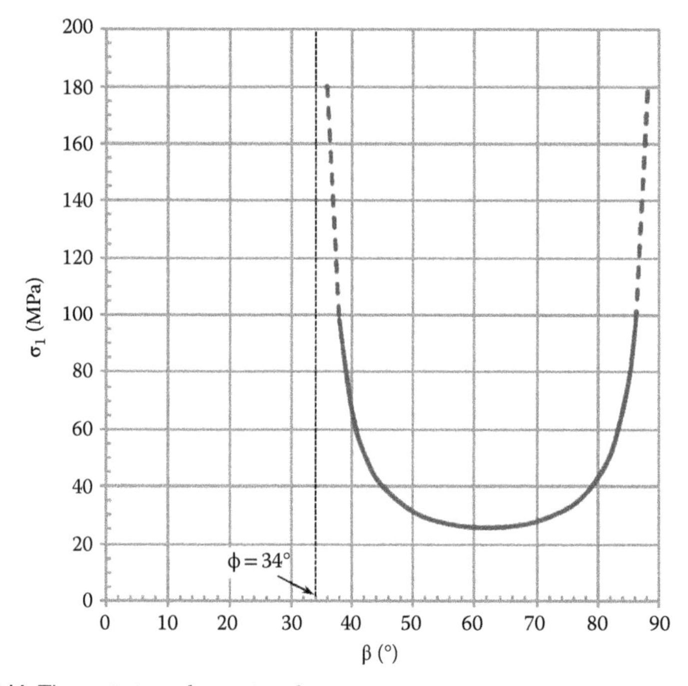

Figure 5.16 The variation of σ_1 against β.

5.9 SUMMARY

1. The presence of discontinuities makes the rock mass behave anisotropically. However, when there are too many discontinuities (e.g., joints), the block size is reduced, and with increased degrees of freedom for movement, the rock mass becomes isotropic and acts like soil.
2. In rocks, the horizontal stress is often larger than the vertical stress. This is opposite to what we see in soils.
3. Rock overburden pressure can be estimated using an average unit weight of 27 kN/m^3.
4. The isotropic linear elastic constitutive model is the simplest of all the models presented. The stresses and the strains can be related by two constants: Young's modulus E and Poisson's ratio ν.
5. Plane strain, plane stress and axisymmetric loadings are three special situations that we encounter when solving engineering problems.
6. The Mohr–Coulomb failure criterion is freely used for soils. In rocks, it does not work very well in the tensile region. σ_c, σ_t and σ_t' can be related to c and ϕ based on the Mohr–Coulomb criterion.
7. The peak shear strength can be significantly larger than the residual one ($\phi_p > \phi_r$, $c_r \approx 0$).
8. The Hoek–Brown failure criterion is more popular than the Mohr–Coulomb for rock mechanics applications. It can be applied to the intact rock as well as the rock mass.
9. The Hoek–Brown constant m_i is analogous to the friction angle ϕ in the Mohr–Coulomb failure criterion. The constant s is analogous to cohesion.
10. The Hoek–Brown constant m_i of an intact rock is approximately equal to the ratio σ_{ci}/σ_{ti}.
11. Typical values: $m = 0$ (weak) to 35 (strong); $s = 0$ (weak) to 1 (strong); $a = 0.50$ (strong) to 0.65 (weak).
12. In the Hoek–Brown model, the parameters for rock mass and the intact rock are related through GSI, which accounts for the quality (interlocking of the blocks and joint surface) of the rock mass.
13. When the failure envelope is drawn on σ_1–σ_3 space, the intercepts of the failure envelope on the two axes give the uniaxial compressive and tensile strengths. This is true for both Mohr–Coulomb and Hoek–Brown failure criteria (see Figure 5.7).
14. For massive rock mass with widely spaced discontinuities, with GSI or RMR approaching 100, the rock mass will have the same strength and modulus as the intact rock.

Review Exercises

1. State whether the following are true or false.
   a. In rocks, K_0 is larger at shallower depths.

b. Generally in rocks, the horizontal stress is greater than the vertical stress.
c. The tensile strength of an intact rock is greater than its Brazilian indirect tensile strength.
d. In the Hoek–Brown failure criterion, the larger the m, the larger is the strength.
e. m_i is always greater than m_m.
f. The larger the GSI, the larger is the strength of a rock mass.

2. What are the non-zero stress components in (a) plane strain loading and (b) plane stress loading?
3. What are the non-zero strain components in (a) plane strain loading and (b) plane stress loading?
4. Carry out a literature survey and list the empirical equations relating σ_{cm}/σ_{ci} to (a) RMR and (b) Q.
5. In plane strain loading, show that the principal strains in terms of principal stresses are given by

$$\sigma_1 = \frac{E}{(1+\nu)(1-2\nu)}\{(1-\nu)\varepsilon_1 + \nu\varepsilon_2\}$$

$$\sigma_3 = \frac{E}{(1+\nu)(1-2\nu)}\{\nu\varepsilon_1 + (1-\nu)\varepsilon_2\}$$

6. Using Equation 5.6, show that for one-dimensional consolidation, the normal stress and the normal strain are related by $\sigma = D\varepsilon$, where D is the *constrained modulus* given by

$$D = \frac{E(1-\nu)}{(1+\nu)(1-2\nu)}$$

7. In a plane strain situation, express the strains in terms of displacements. The square object shown in the following figure is subjected to plane strain loading where the displacements u and ν are given by $u = x^2y$ and $\nu = xy^3$. Determine the strain components in terms of x and y.

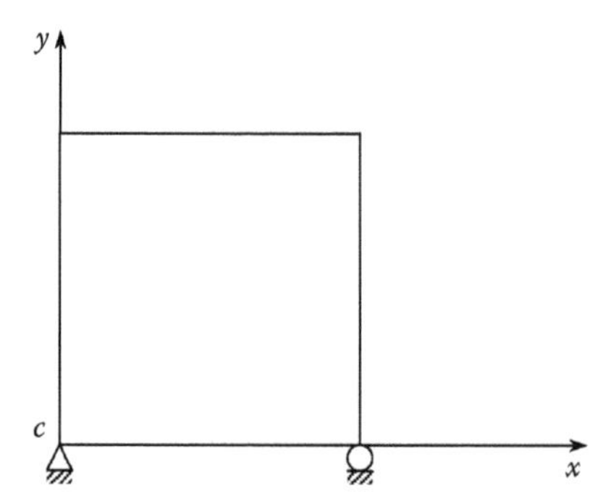

8. The same square object shown in the figure in the previous exercise is subjected to plane strain loading where the strains are given by $\varepsilon_x = 2xy$, $\varepsilon_y = 3xy^2$ and $\gamma_{xy} = x^2 + y^3$. Applying appropriate boundary conditions, develop the expressions for the displacements.
9. Based on the Mohr–Coulomb failure criterion and assuming that it holds in the tensile region as well, what is the ratio of tensile strength to indirect tensile strength for typical values of friction angle?
   Answer: 1–2
10. Triaxial tests were carried out on 54-mm-diameter (NX core) intact rock specimens. The applied confining pressures and the principal stress differences at failure are summarised below.

Confining pressure (MPa)	0	5.0	10.0	15.0	20.0	25.0
Principal stress difference (MPa)	59.5	87.5	116.0	139.5	167.5	192.5

    Plot σ_1 against σ_3 at failure and determine the uniaxial compressive strength and the friction angle of the intact rock.
11. The intact rock that follows the Mohr–Coulomb failure criterion has c = 15 MPa and ϕ = 27°. Estimate the uniaxial compressive strength σ_c, the uniaxial tensile strength σ_t and the Brazilian indirect tensile strength σ_t'. Give an estimate of the point load strength index $I_{s(50)}$ and suggest a realistic range for the Young's modulus E.
    Answer: 49.0 MPa, 18.4 MPa, 8.6 MPa, 2.0 MPa and 14.7–49.0 GPa
12. The rock mass at a hydroelectric powerhouse project in Himachal Pradesh, India, consists of jointed quartz mica schist with average GSI of about 65 (Hoek and Brown, 1997). Triaxial tests on intact rock cores showed σ_{ci} = 30 MPa and m_i = 15.6.
    a. Estimate the rock mass parameters m_m, s and a.
    b. Estimate c' and ϕ' of the rock mass.
    c. Estimate the uniaxial compressive and tensile strengths of the rock mass.

    Answer: 4.47, 0.021, 0.502 and 1.9 MPa, 39°
13. Using the values in Exercise 12 (i.e., σ_{ci} = 30 MPa, m_m = 4.47, s = 0.021 and a = 0.502) in the generalised Hoek–Bray failure criterion for rock mass, generate the values of σ_{1f}' corresponding to σ_{3f}' = 0, 2.5, 5, 7.5, 10, 12.5 MPa.
14. Undisturbed specimens of the gouge material filling a rock joint was tested in the laboratory and the cohesion and friction angles are determined as 5 MPa and 35°, respectively. If the minor principal stress at the joint is 2 MPa, determine the value of σ_1 that is required to cause shear failure along the joint that is inclined to the major principal plane by (a) 45°, (b) 55° and (c) 65°.
    Answer: 44.7, 28.7 and 26.8 MPa

REFERENCES

Bieniawski, Z.T. (1978). Determining rock mass deformability—Experience from case histories. *International Journal of Rock Mechanics and Mining Sciences*, Vol. 15, No. 5, pp. 237–248.

Goodman, R.E. (1989). *Introduction to Rock Mechanics*. 2nd edition, Wiley, New York.

Grimstad, E. and Barton, N. (1993). Updating the Q-system for NMT. *Proceedings of the International Symposium on Sprayed Concrete*, Fagernes, Norway, Norwegian Concrete Association, Oslo, 20pp.

Hoek, E. (1983). Strength of jointed rock masses, 23rd Rankine lecture. *Geotechnique*, Vol. 33, No. 3, pp. 187–223.

Hoek, E. (2007). *Practical Rock Engineering*. http://www.rocscience.com/hoek/corner/practical_rock_engineering.pdf.

Hoek, E. and Brown, E.T. (1980a). Empirical strength criterion for rock masses. *Journal of Geotechnical Engineering Division*, ASCE, Vol. 106, No. GT9, pp. 1013–1035.

Hoek, E. and Brown, E.T. (1980b). *Underground Excavation in Rock*. Institute of Mining and Metallurgy, London.

Hoek, E. and Brown, E.T. (1997). Practical estimates of rock mass strength. *International Journal of Rock Mechanics and Mining Sciences*, Vol. 34, No. 8, pp. 1165–1186.

Hoek, E., Carranza-Torres, C.T. and Corkum, B. (2002). Hoek-Brown failure criterion – 2002 edition. *Proceedings of the 5th North American Rock Mechanics Symposium*, Toronto, Canada, Vol. 1, pp. 267–273.

Hoek, E. and Diederichs, M.S. (2006). Empirical estimation of rock mass modulus. *International Journal of Rock Mechanics and Mining Sciences*, Vol. 43, No. 2, pp. 203–215.

Hondros, G. (1959). The evaluation of Poisson's ratio and the modulus of materials of a low tensile resistance by the Brazilian (indirect tensile) test with particular reference to concrete. *Australian Journal of Applied Science*, Vol. 10, No. 3, pp. 243–268.

Marinos, P. and Hoek, E. (2001). GSI: A geologically friendly tool for rock mass strength estimation. *Proceedings of the International Conference on Geotechnical & Geological Engineering, GeoEng2000*, Technomic Publications, Melbourne, Australia, 1422–1442.

Shorey, P.R. (1994). A theory for in situ stresses in isotropic and transversely isotropic rock. *International Journal of Rock Mechanics and Mining Sciences & Geomechanics Abstracts*, Vol. 31, No. 1, pp. 23–34.

Serafim, J.L. and Pereira, J.P. (1983). Consideration of the geomechanical classification of Bieniawski. *Proceedings of the International Symposium on Engineering Geology and Underground Construction*, Lisbon, 1(II), pp. 33–44.

Chapter 6

Rock slope stability

6.1 INTRODUCTION

Rock slopes either occur naturally (Figure 6.1a) or are engineered by people as products of excavations to create space for buildings, highways and railway tracks, powerhouses, dams and mine pits (Figure 6.1b). The analysis for the estimation of stability of rock slopes has been a challenging task for engineers, especially under hydraulic and seismic conditions. In most civil and mining engineering projects, the main purpose of slope stability analysis is to contribute to the safe and economic design of rock slopes. This chapter describes the basic modes/mechanisms of rock slope failures and presents the fundamental concepts and methods of rock slope stability analysis. In field situations, many rock slopes are unstable, or they require an improvement in their stability. Such slopes need to be stabilised as per the specific needs of the project. Therefore, this chapter introduces some common rock slope stabilisation techniques.

6.2 MODES OF ROCK SLOPE FAILURE

The modes of rock slope failure depend mainly on the geometric interaction of existing discontinuities (jointing and bedding patterns) and free space/excavation surfaces in the rock mass constituting the slope. For safe and economic design of rock slopes, it is important to recognise the modes/mechanisms in which slopes in rock masses can fail. This task requires good engineering judgment, which can be achieved by good engineering practice that deals with rock slopes in varied geologic terrain. The spherical presentation of geological data (dip and strike) helps identify the most likely basic potential modes of rock slope failure (see Section 2.6). The measurement of piezometric levels and springs throughout the slope, and measurements of slope deformations (with slope inclinometers and precise surveying of fixed surficial targets) are other basic tools to judge the most likely potential failure modes of rock slope failure. The idealised, simple,

(a)

(b)

Figure 6.1 Rock slopes: (a) natural rock slope; (b) engineered (excavated) rock slope.

basic modes of rock slope failure that are considered in practice are the following (Hoek and Bray, 1981; Goodman, 1989; Goodman and Kieffer, 2000; Wyllie and Mah, 2004):

1. Plane failure
2. Wedge failure
3. Circular failure
4. Toppling failure

In plane failure mode, the rock block slides on a single face that can be a joint plane or bedding plane striking parallel to the slope face and dipping into free space/excavation at an angle greater than the angle of internal friction of the joint/bedding material (Figure 6.2a). In wedge failure mode, the wedge of rock slides simultaneously on two discontinuity planes, striking obliquely across the slope face, along their line of intersection daylighting into the slope face, provided that the inclination of this line is significantly

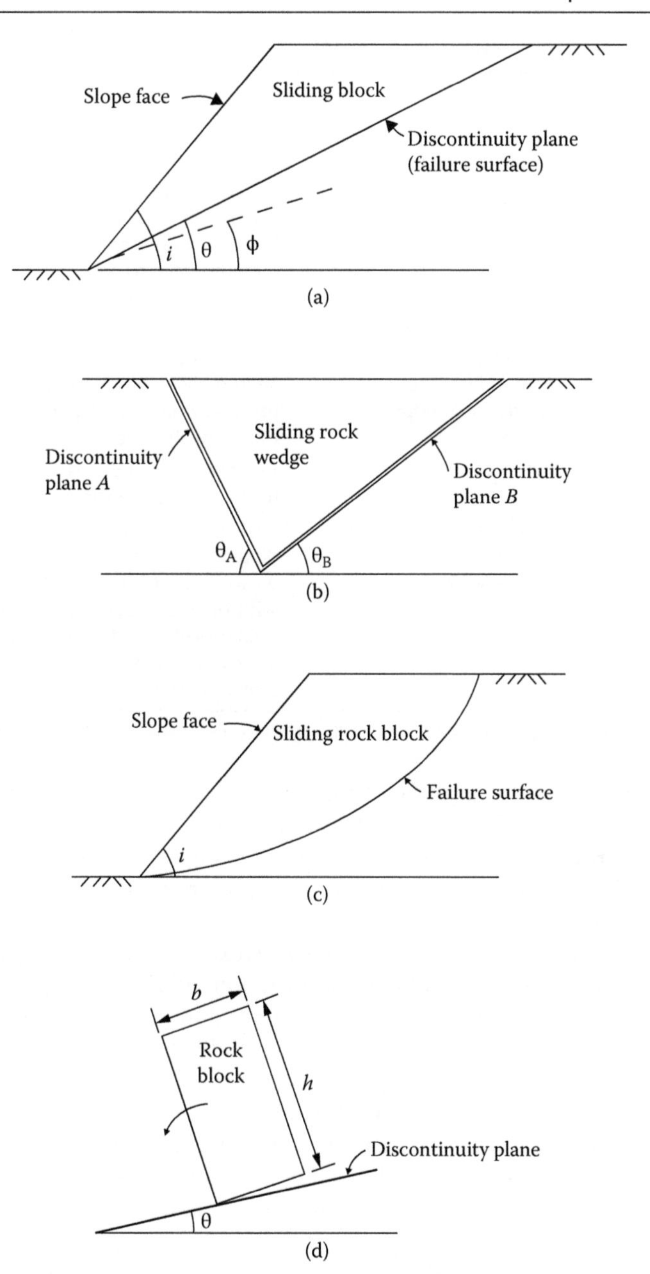

Figure 6.2 Basic modes of rock slope failure: (a) plane failure ($i > \theta > \phi$; i is slope inclination to the horizontal, θ is inclination of the discontinuity plane/failure plane and ϕ is angle of internal friction of the joint/bedding material); (b) wedge failure (θ_A and θ_B are the inclinations of discontinuity planes A and B, respectively); (c) circular failure; (d) toppling failure (b is width and h is height of the rock block).

Table 6.1 Basic modes of rock slope failure

Mode of rock failure	*Description*	*Typical materials*
Plane failure	Sliding without rotation along a face; single or multiple blocks	Hard or soft rocks with well-defined discontinuities and jointing, e.g. layered sedimentary rocks, volcanic flow rocks, block-jointed granite, foliated metamorphic rocks
Wedge failure	Sliding without rotation on two nonparallel planes, parallel to their line of intersection; single or multiple blocks	Blocky rocks with at least two continuous and nonparallel joint sets, e.g. cross-jointed sedimentary rocks, regularly faulted rocks, block-jointed granite and especially foliated or jointed metamorphic rocks
Circular failure	Sliding on a cylindrical face	Heavily jointed and weathered rock masses similar to the soils
Toppling failure	Forward rotation about an edge/base; single or multiple blocks	Hard rocks with regular, parallel joints dipping away from the free space/excavation, i.e. dipping into the hillside, with or without crossing joints; foliated metamorphic rocks and steeply dipping layered sedimentary rocks; also in block-jointed granites

Source: Adapted from Goodman, R.E., and Kieffer, D.S., *J. Geotech. Geoenviron. Eng.*, 126, 675–684, 2000.

greater than the average angle of internal friction of the two joint/bedding materials (Figure 6.2b). In circular failure mode, the heavily jointed and weathered rock mass, similar to a waste dump rock, slides on a single cylindrical face into free space/excavation (Figure 6.2c). In toppling failure mode, the multiple rock columns/layers caused by a steeply dipping joint set rotate about their bases into the free space/excavation (Figure 6.2d). Plane and wedge failures are more common than circular and toppling failures. Toppling failure can be very significant, if not dominant, in some rock types of steep mountain slopes or open pit mines. Table 6.1 describes these failure modes, and gives examples of typical materials in which they are realised.

It should be noted that if a rock slope is large and embraces a mix of rock types and structures, more than one of the basic failure modes may be expected. On the contrary, within a single sliding mass, it is not unreasonable to find more than one of the basic failure modes at the site.

6.3 SLOPE STABILITY ANALYSIS

The stability of rock slopes is greatly controlled by the shear strength along the joints and interfaces between the unstable rock block/wedge and intact rock, as well as by the geometric interaction of jointing and bedding patterns in the rock mass constituting the slope. The magnitude of the available shear strength along joints and interfaces is very difficult to determine due to the inherent variability of the material and the difficulties associated with sampling and laboratory testing. Depending on the critical nature of the project, field direct shear tests are performed on joints in an effort to determine reliable strength parameters. Factors that directly or indirectly influence the strength include the following (Bromhead, 1992; Abramson et al., 2002):

1. The planarity and smoothness of the joint's surfaces. A smooth planar surface will have a lower strength than an irregular and rough surface.
2. The inclination of the discontinuity plane with respect to the slope.
3. The openness of the discontinuity, which can range from a small fissure to a readily visible joint.
4. The extent of the weathering along the surfaces and the possible infill of the joint with weaker material such as clays and calcareous materials. A calcareous infill may potentially increase the strength of the joint, whereas a soft clay infill may reduce the strength of the joint to the same level as the clay material itself. Such infills may also change the seepage pattern, improving or degrading the drainage, which will be manifested by an increase or decrease in pore water pressures within the joints.

Once the failure modes have been recognised and the joint strengths have been determined, the factor of safety can be estimated using the principles of statics, with free-body diagrams deduced from the geological map that describes the geological structures, and water/seepage forces calculated from the piezometric measurements. Limit equilibrium methods have been useful in developing the fundamental understanding of rock slope stability analysis for simple modes of failure. Numerical methods help analyze the rock slopes, especially failing in a combination of basic modes and/or other known failure modes (erosion, ravelling, slumping, block torsion, sheet failure, buckling, bursting etc., as listed by Goodman and Kieffer [2000]). We discuss here the fundamentals of limit equilibrium methods of rock slope stability analysis, and present analytical expressions for simple static and seismic loading conditions.

6.3.1 Factor of safety

The task of the engineer analyzing the stability of a rock slope is to determine the factor of safety. In general, the factor of safety (FS) against sliding of a rock block is defined as

$$\mathrm{FS} = \frac{F_\mathrm{r}}{F_\mathrm{i}} \tag{6.1}$$

where F_r is the total force available to resist the sliding of the rock block and F_i is the total force tending to induce sliding. For a slope on the point of failure, a condition of limiting equilibrium exists in which $F_r = F_i$, and thus FS = 1. For stable slopes, $F_r > F_i$, and therefore FS > 1. In practice, rock slopes with FS = 1.3 to 1.5 are considered to be stable; the lower value is taken for temporary slopes such as mine slopes, whereas the higher value is considered for permanent slopes such as slopes adjacent to road pavements and railway tracks.

6.3.2 Plane failure

Figure 6.3 shows a rock slope of height H inclined to the horizontal at an angle i. The sliding rock block $A_1A_2A_3$ is separated by the joint/bedding/failure plane A_2A_3, which is inclined to the horizontal at an angle θ. A_1A_3 (= B) is the top width of the sliding rock block, and W is its weight. The stability of the rock block $A_1A_2A_3$ is analysed as a two-dimensional limit equilibrium problem, considering a slice of unit thickness through the slope. Only the force equilibrium is considered, neglecting any resistance to sliding at the lateral boundaries of the sliding block. The joint/bedding plane material is assumed to be a c–ϕ soil material, with c and ϕ as cohesion and angle of internal friction (also called angle of shearing resistance), respectively, obeying the Mohr–Coulomb failure criterion.

The total force available to resist the sliding block is

$$F_r = sA \tag{6.2}$$

where s is the shear strength of the sliding failure plane, and A is the area of the base A_2A_3 of the sliding rock block. It is given as

$$A = \frac{H}{\sin\theta} \tag{6.3}$$

The top width B is calculated as

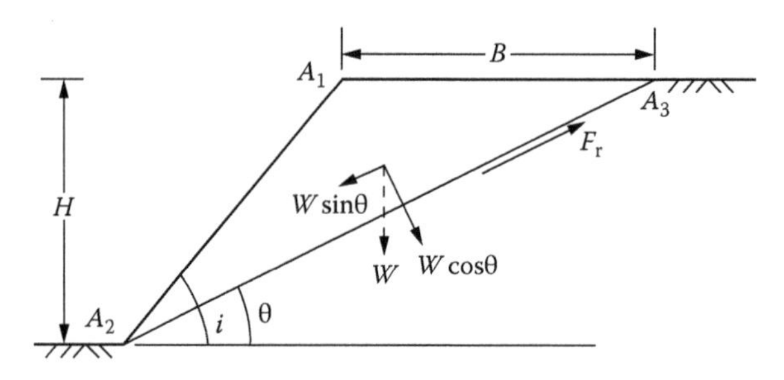

Figure 6.3 A rock slope in plane failure.

$$B = H(\cot\theta - \cot i) = \frac{H\sin(i-\theta)}{\sin i \sin\theta} \tag{6.4}$$

The Mohr–Coulomb failure criterion provides

$$s = c + \sigma_n \tan\phi \tag{6.5}$$

where σ_n is the normal stress on the failure plane. From Equations 6.2 and 6.5,

$$F_r = cA + F_n \tan\phi \tag{6.6}$$

where $F_n = \sigma_n A$ is the normal force on the failure plane. Considering equilibrium of forces acting on the rock block in a direction normal to the slope face, F_n is obtained as

$$F_n = W\cos\theta \tag{6.7}$$

The weight W is calculated as

$$W = \frac{1}{2}\gamma BH$$

or, using Equation 6.4, we get

$$W = \frac{1}{2}\left[\frac{\sin(i-\theta)}{\sin i \sin\theta}\right]\gamma H^2 \tag{6.8}$$

Substituting values from Equations 6.3, 6.7 and 6.8 into Equation 6.6, we arrive at

$$F_r = \frac{cH}{\sin\theta} + \frac{1}{2}\left[\frac{\sin(i-\theta)\cos\theta}{\sin i \sin\theta}\right]\gamma H^2 \tan\phi \tag{6.9}$$

From Figure 6.3, the total force tending to induce sliding is calculated as

$$F_i = W\sin\theta$$

or, using Equation 6.8, we get

$$F_i = \frac{1}{2}\left[\frac{\sin(i-\theta)}{\sin i}\right]\gamma H^2 \tag{6.10}$$

Substituting F_r and F_i from Equations 6.9 and 6.10, respectively, into Equation 6.1, the factor of safety is obtained as

$$FS = \frac{2c\sin i}{\gamma H \sin\theta \sin(i-\theta)} + \frac{\tan\phi}{\tan\theta} \tag{6.11}$$

or

$$FS = \frac{2c^{*}\sin i}{\sin\theta \sin(i-\theta)} + \frac{\tan\phi}{\tan\theta} \tag{6.12}$$

where $c^{*} = c/\gamma H$ is a nondimensional parameter that may range between 0 and 1, although c, γ and H vary over wide ranges. From Equation 6.11, it is noted that the factor of safety is a function of six parameters (c, γ, H, i, θ, ϕ), whereas Equation 6.12 states that it is a function of only four parameters (c^{*}, i, θ, ϕ), which are nondimensional. Therefore, Equation 6.12 can be conveniently used for preparing design charts for the design of simple rock slopes against plane failure. The authors recommend calculation of the factor of safety using Equation 6.11 or Equation 6.12 in the MS Excel spreadsheet, in place of using developed design charts or a pocket calculator, to save design time, especially when several rock slopes have to be analysed and designed.

EXAMPLE 6.1

For the rock slope shown in Figure 6.3, consider that the joint/bedding material is cohesionless. What is the expression for the factor of safety? Under what condition can the slope fail?

Solution

For cohesionless joint/bedding material, $c = 0$, and Equation 6.12 reduces to

$$FS = \frac{\tan\phi}{\tan\theta} \tag{6.13}$$

For failure of the slope,

$$FS < 1$$

or

$$\tan\theta > \tan\phi$$

or

$\theta > \phi$; that is, the inclination of the joint/bedding plane to the horizontal should be greater than the angle of internal friction of the joint/bedding material, which has been stated in Section 6.2.

It should be noted that Figure 6.3 presents a simple case of plane failure, which is not a very common field situation; however, this case is very useful in understanding how the variation in basic factors can govern the stability of rock slopes against plane failure. In reality, some or all of the following factors/physical situations can be present at many field sites (Shukla et al., 2009; Hossain and Shukla, 2010; Shukla and Hossain, 2010; Shukla and Hossain, 2011a, b):

- Tension crack in slope with no water
- Tension crack in rock slope filled with water partially or fully
- Seepage pressure at the joint/bedding plane
- Surcharge at the top of the slope
- Horizontal and vertical seismic loads
- Stabilising force through reinforcing system such as rock bolts, anchors and cables

Figure 6.4 shows an anchored rock slope of height H with an inclination i to the horizontal. The joint/bedding plane A_2A_3 inclined to the horizontal at an angle θ and a vertical tension crack A_3A_4 of depth z separate a portion of the rock mass as the block $A_1A_2A_3A_4$ having a weight W. The tension crack is filled with water having a unit weight γ_w to a depth z_w. The

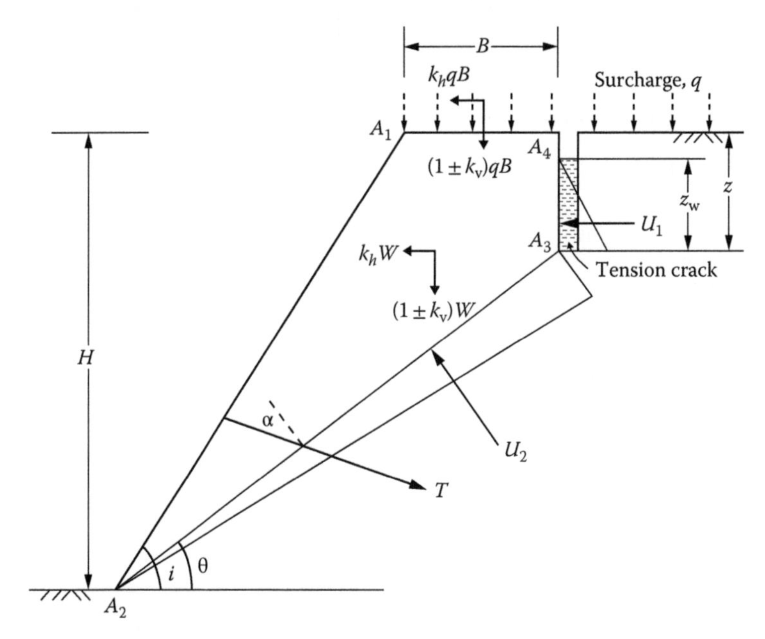

Figure 6.4 Anchored rock slope with a water-filled tension crack subjected to surcharge and seismic loads in plane failure along the joint/bedding plane. (Adapted from Shukla, S.K., and Hossain, M.M., *Int. J. Geotech. Eng.*, 5, 181–187, 2011.)

stabilising tensile force T inclined at an angle α to the normal at the joint/bedding plane A_2A_3 simulates the effect of a rock-anchoring system, which is commonly used to stabilise rock slopes. The horizontal and vertical seismic inertial forces, k_hW and k_vW with k_h and k_v [↓+ and ↑−] as horizontal and vertical seismic coefficients, respectively, are shown to act on the sliding block. A surcharge placed at the top of the slope A_1A_4 (= B) applies a vertical pressure q along with horizontal and vertical seismic inertial forces, k_hqB and k_vqB, respectively. The horizontal force due to water pressure in the tension crack is U_1. The water in the tension crack seeps through the joint/bedding plane and applies an uplift force U_2. Under a critical combination of forces, the rock mass block $A_1A_2A_3A_4$ can slide along the joint/bedding plane A_2A_3 as a failure plane.

The expression for the factor of safety of the slope shown in Figure 6.4 against plane failure can be derived by following the steps described for the simple slope shown in Figure 6.3. The readers can find the complete derivation in the research article by Shukla and Hossain (2011a), where the factor of safety is given as

$$\text{FS} = \frac{2c^*P + \left[(1 \pm k_v)(Q + 2q^*R)\left\{\dfrac{\cos(\theta+\psi)}{\cos\psi}\right\} - \left(\dfrac{z_w^{*2}}{\gamma^*}\right)\sin\theta - \left(\dfrac{z_w^*}{\gamma^*}\right)P + 2T^*\cos\alpha\right]\tan\phi}{(1 \pm k_v)(Q + 2q^*R)\left\{\dfrac{\sin(\theta+\psi)}{\cos\psi}\right\} + \left(\dfrac{z_w^{*2}}{\gamma^*}\right)\cos\theta - 2T^*\sin\alpha} \tag{6.14}$$

in terms of the following nondimensional parameters:

$$c^* = \frac{c}{\gamma H} \tag{6.15a}$$

$$z^* = \frac{z}{H} \tag{6.15b}$$

$$z_w^* = \frac{z_w}{H} \tag{6.15c}$$

$$\gamma^* = \frac{\gamma}{\gamma_w} \tag{6.15d}$$

$$q^* = \frac{q}{\gamma H} \tag{6.15e}$$

$$T^* = \frac{T}{\gamma H^2} \tag{6.15f}$$

$$P = (1 - z^*)\operatorname{cosec}\theta \tag{6.15g}$$

$$Q = (1 - z^{*2})\cot\theta - \cot i \tag{6.15h}$$

$$R = (1 - z^*)\cot\theta - \cot i \tag{6.15i}$$

and

$$\psi = \tan^{-1}\left(\frac{k_h}{1 \pm k_v}\right) \tag{6.15j}$$

Equation 6.14 is a general expression for the factor of safety of the rock slope against plane failure. It can be used to investigate the effect of any individual parameter on the factor of safety of the rock slope and to carry out a detailed parametric study as required in a specific field situation. There can be several special cases of Equation 6.14, including expressions in Equations 6.12 and 6.13, and many of them have been presented in the literature (Hoek and Bray, 1981; Ling and Cheng, 1997; Wyllie and Mah, 2004; Shukla and Hossain, 2011a, b).

Seismic coefficients k_h and k_v are expressed as fractions of the gravitational constant. In conventional pseudostatic methods of analysis, the choice of horizontal seismic coefficient, k_h, for design is related to the specified horizontal peak ground acceleration for the site, a_h. The relationship between a_h and a representative value of k_h is, nevertheless, complex, and there does not appear to be a general consensus in the literature on how to relate these parameters. Values of k_h from 0.05 to 0.15 are typical for design, and these values correspond to 1/3 to 1/2 of the peak acceleration of the design earthquake (Bathurst et al., 2012). In practice, the choice of k_h should be based on local experience or prescribed by local building codes or other regulations. The experience suggests that k_h may be as high as 0.5, and k_v is generally taken as half of k_h.

EXAMPLE 6.2

For the rock slope shown in Figure 6.4, consider the following: $i = 50°$, $\theta = 35°$, $\phi = 25°$, $q^* = 0.5$, $T^* = 0.1$, $z^* = 0.1$, $z_w^* = 0.05$, $\gamma^* = 2.5$, $\alpha = 45°$ and $c^* = 0.1$. Plot the variation of the factor of safety (FS) with vertical seismic coefficient (k_v) for horizontal seismic coefficient, $k_h = 0.05$, 0.1, 0.15, 0.2, 0.25 and 0.3. Assume that the maximum value of $k_v = k_h/2$. What do you notice in this plot?

Solution

Using Equation 6.14, the variation of factor of safety (FS) with vertical seismic coefficient (k_v) for the given values of horizontal seismic coefficient (k_h) is shown in Figure 6.5.

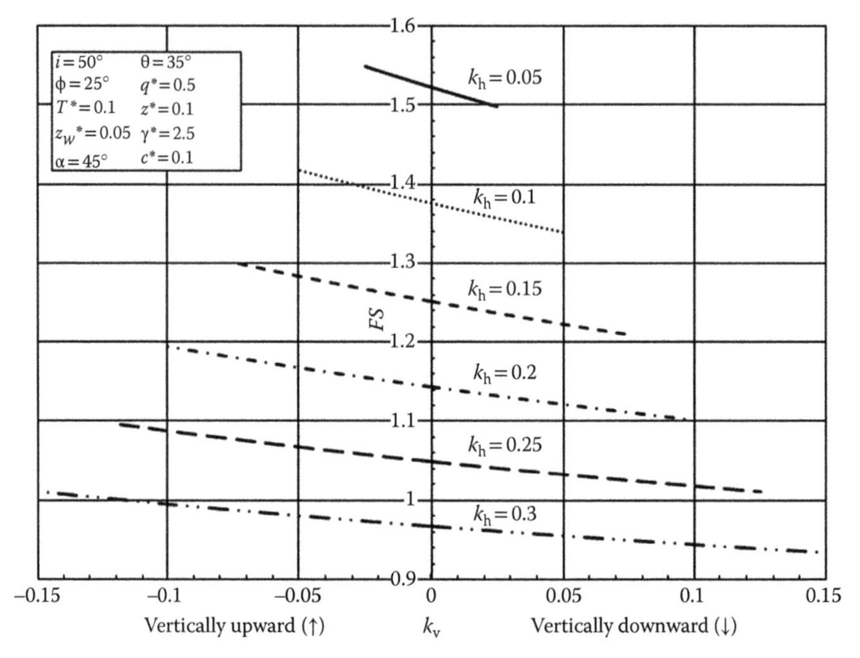

Figure 6.5 Variation of factor of safety (FS) with vertical seismic coefficient k_v.

The following two key observations are noted:

1. With an increase in k_v in the downward direction, FS decreases almost linearly, but it increases as k_v increases in the upward direction.
2. FS is greater than unity for any value of k_h less than 0.25, and it is higher for smaller values of k_h, which is an expected observation.

Following the graphical approach adopted in Example 6.2, Equation 6.14 can be used to develop design charts for specific field parameters. Shukla and Hossain (2010) have presented examples of some design charts for assessing the stability of anchored rock slopes against plane failure. Figure 6.6 shows a typical design chart.

6.3.3 Wedge failure

Figure 6.7 shows forces acting on a rock wedge $A_1A_2A_3$ in its two views: (a) view looking at the wedge face and (b) cross-sectional view. R_A and R_B are the normal reactions provided to the sliding wedge by planes A and B, respectively. A condition of wedge sliding is defined by $i > \beta > \phi_{av}$, where i is the slope face inclination to the horizontal as considered in plane failure, β is the inclination to the horizontal of the line (i.e. plunge) of intersection of discontinuity planes A and B and ϕ_{av} is the average angle of internal friction

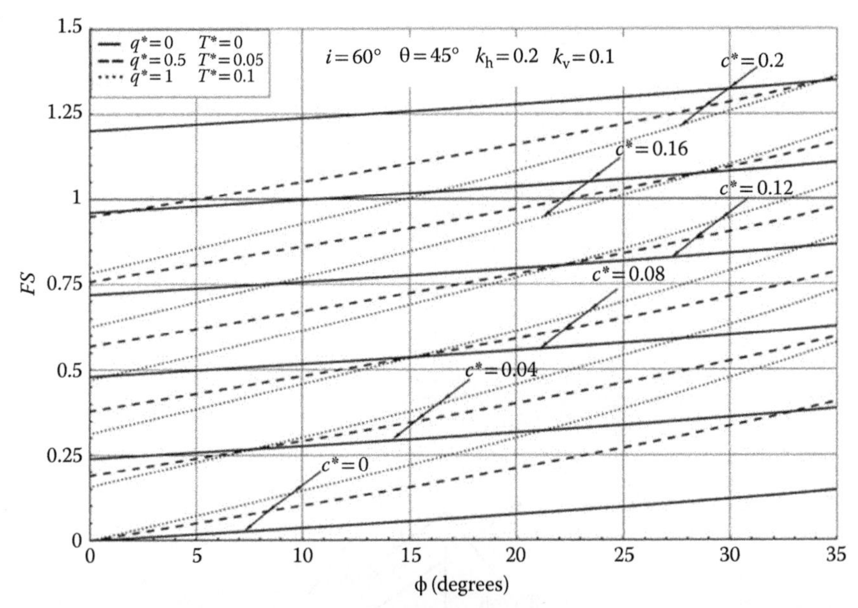

Figure 6.6 A typical design chart for estimating the stability of anchored rock slopes against plane failure. (Adapted from Shukla, S.K., and Hossain, M.M., Design charts for rock slope stability against plane failure under seismic loading condition. *Proceedings of the ISRM International Symposium 2010 and 6th Asian Rock Mechanics Symposium*, October 23–27, 2010, New Delhi, India, Paper No. 64, 2010.)

for the two discontinuity/slide planes A and B. If the angle of internal friction is the same for both planes and is ϕ, ϕ_{av} will be equal to ϕ. The cohesive forces at the discontinuity planes are assumed to be negligible.

The total force available to resist the sliding of the rock wedge along the line of intersection is

$$F_r = R_A \tan\phi + R_B \tan\phi = (R_A + R_B)\tan\phi \tag{6.16}$$

The total force tending to induce sliding along the line of intersection is

$$F_i = W \sin\beta \tag{6.17}$$

Substituting F_r and F_i from Equations 6.16 and 6.17, respectively, into Equation 6.1, the factor of safety is obtained as

$$\text{FS} = \frac{(R_A + R_B)\tan\phi}{W \sin\beta} \tag{6.18}$$

Resolving forces R_A and R_B into components normal and parallel to the direction along the line of intersection, we get

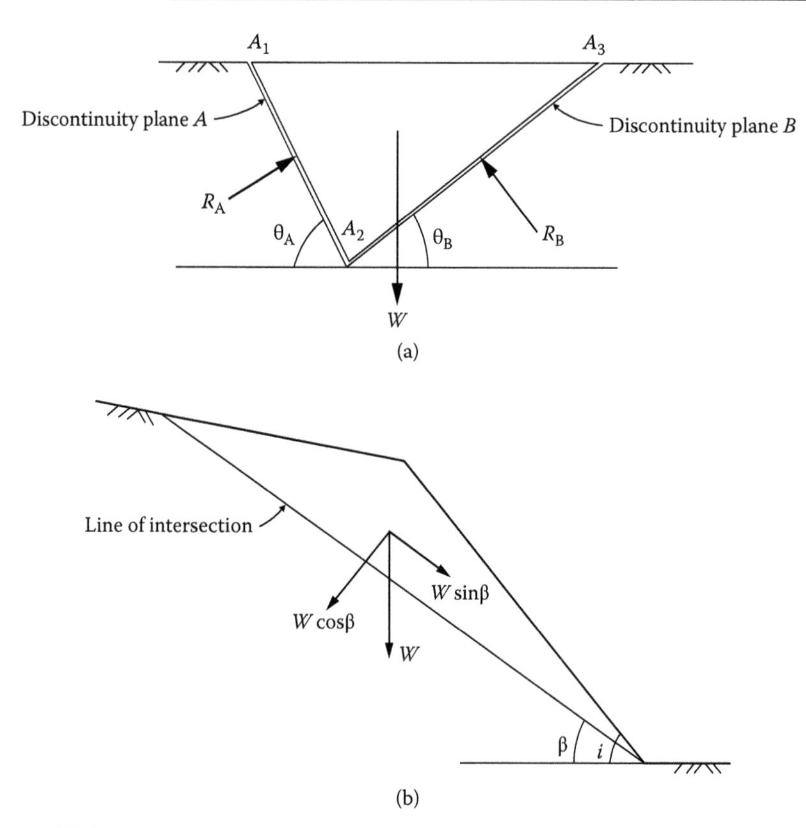

Figure 6.7 Forces acting on the rock wedge: (a) view of wedge looking at its face showing definition of angles θ_A and θ_B, and reactions R_A and R_B of discontinuity planes A and B, respectively; (b) cross section of wedge showing resolution of the weight W.

$$R_A \sin\theta_A = R_B \sin\theta_B \tag{6.19}$$

and

$$R_A \cos\theta_A + R_B \cos\theta_B = W \cos\beta \tag{6.20}$$

Solving Equations 6.19 and 6.20 for R_A and R_B, we obtain

$$R_A = \frac{W \cos\beta \sin\theta_B}{\sin(\theta_A + \theta_B)} \tag{6.21}$$

and

$$R_B = \frac{W \cos\beta \sin\theta_A}{\sin(\theta_A + \theta_B)} \tag{6.22}$$

Using Equations 6.21 and 6.22, Equation 6.18 can be expressed as

$$FS = \frac{(\sin\theta_A + \sin\theta_B)\tan\phi}{\tan\beta\sin(\theta_A + \theta_B)} \tag{6.23}$$

or

$$FS = K\left(\frac{\tan\phi}{\tan\beta}\right) \tag{6.24}$$

where

$$K = \frac{\sin\theta_A + \sin\theta_B}{\sin(\theta_A + \theta_B)} \tag{6.25}$$

K is a wedge factor that depends on the inclinations of the discontinuity planes and is greater than 1. If the factor of safety FS against the wedge failure is denoted by FS_W, Equation 6.24 can be written as

$$FS_W = K(FS_P) \tag{6.26}$$

where FS_P (= $\tan\phi/\tan\beta$) is the factor of safety of the rock slope against plane failure in which the slide plane, with an angle of internal friction ϕ, dips at the same angle β as the line of intersection of the planes A and B.

The wedge failure analysis presented here does not incorporate different friction angles and cohesions on the two slide planes, groundwater seepage, surcharge and seismic loads. When these factors are included in the analysis, the analytical expressions become complex; more details are presented by Hoek and Bray (1981) and Wyllie and Mah (2004).

EXAMPLE 6.3

For the rock slope shown in Figure 6.7, consider the following: $i = 62°$, $\beta = 53°$, $\phi = 30°$, $\theta_A = 45°$ and $\theta_B = 48°$. Calculate the factor of safety. Is the slope stable?

Solution

From Equation (6.25), $K = \dfrac{\sin 45° + \sin 48°}{\sin(45° + 48°)} = 1.45$

From Equation (6.24), $FS = 1.45\left(\dfrac{\tan 30°}{\tan 53°}\right) = 0.63$

Since FS < 1, the slope is unstable.

6.3.4 Circular failure

In the case of a closely jointed/fractured and highly weathered rock slope, the slide surface is free to find the line of least resistance through the slope. In such materials, it is observed that the slide surface generally takes the form of a cylindrical surface that has a circular cross section; therefore, the failure is called circular failure (Figure 6.1c), which is the most common type of slope failure in soils. Various methods of analysis for circular failure in soils have been described in detail in textbooks dealing with soil mechanics; the readers may refer to Taylor (1948), Terzaghi (1943), Lambe and Whitman (1979), Terzaghi et al. (1996) and Das (2013).

6.3.5 Toppling failure

Toppling failures occur in a wide range of rock masses in both natural and engineered slopes. They involve the rotation of columns or blocks of rocks about their bases. The simplest toppling mechanisms involve a single block, resulting in single-block toppling or flexural toppling as illustrated in Figure 6.8. The former mode of toppling occurs when the rock block is already detached from the rock mass of the slope, and the latter occurs

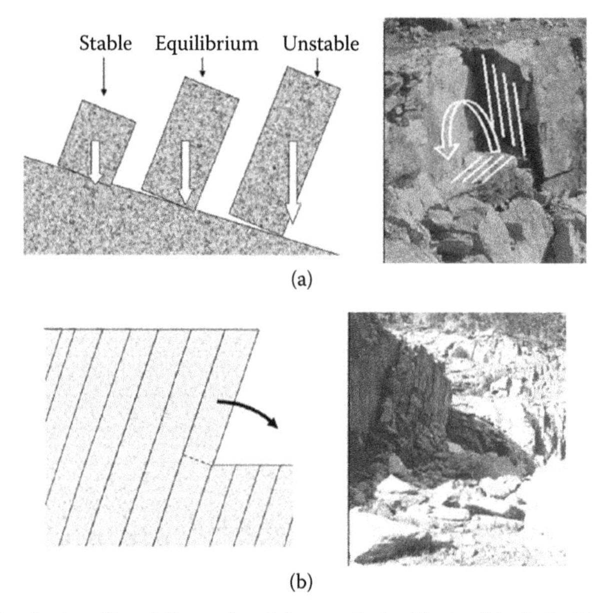

Figure 6.8 Simple toppling failures involving a single block: (a) single-block toppling; (b) single-block flexural toppling. (After Alejano, L.R., et al., *Eng. Geol.*, 114, 93–104, 2010.)

when the rock block remains attached to the rock mass of the slope. The most common toppling failures involve several blocks, and they can be classified as (Figure 6.9) block toppling, flexural toppling and block–flexural toppling (Hoek and Bray, 1981; Goodman and Kieffer, 2000; Wyllie and Mah, 2004). Block toppling takes place in a hard rock mass when individual

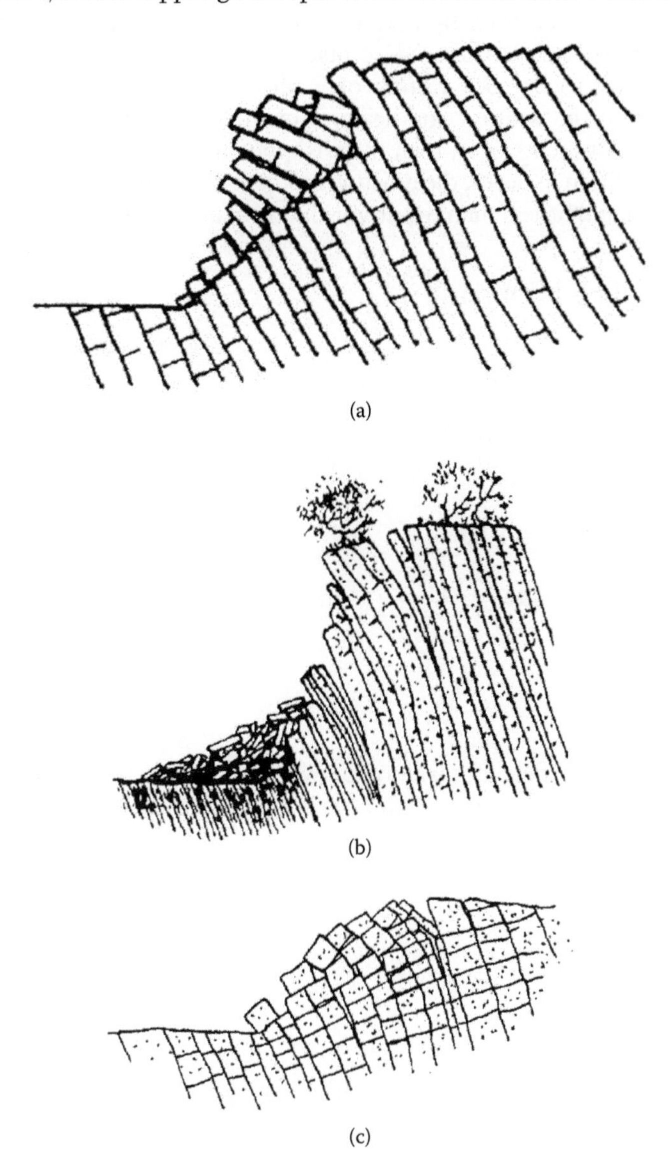

Figure 6.9 Common toppling failures involving several rock blocks: (a) block toppling; (b) flexural toppling; (c) block–flexural toppling. (From Goodman, R.E., and Kieffer, D.S., *J. Geotech. Geoenviron. Eng.*, 126, 675–684, 2000.)

blocks or columns are composed of two normal joint sets, with the main set dipping steeply into the slope face. The upper blocks tend to topple, and push forward on the short columns in the slope toe. Flexural toppling occurs when continuous columns of rock dipping steeply towards the slope break in flexure and tilt forward. Block–flexural toppling is a complex mechanism characterised by pseudocontinuous flexure along blocks that are divided by a number of cross-joints.

For a single-rock block resting on a discontinuity plane, as shown in Figure 6.10, if the width b and height h of the rock block are such that its weight acts outside its base, then there is a potential for the block to topple. For this condition to occur, the resisting moment about the outer lower edge of the block should be less than the driving moment about the same edge; that is,

$$(W\cos\theta)\left(\frac{b}{2}\right) < (W\sin\theta)\left(\frac{h}{2}\right)$$

or

$$\frac{b}{h} < \tan\theta \qquad (6.27)$$

For the sliding of the block,

$$W\sin\theta > \mu W\cos\theta$$

or

$$\tan\theta > \mu \qquad (6.28)$$

where μ is the coefficient of friction between the sliding block and the joint/bedding plane. Since $\mu = \tan\phi$, inequality 6.28 becomes

$$\theta > \phi \qquad (6.29)$$

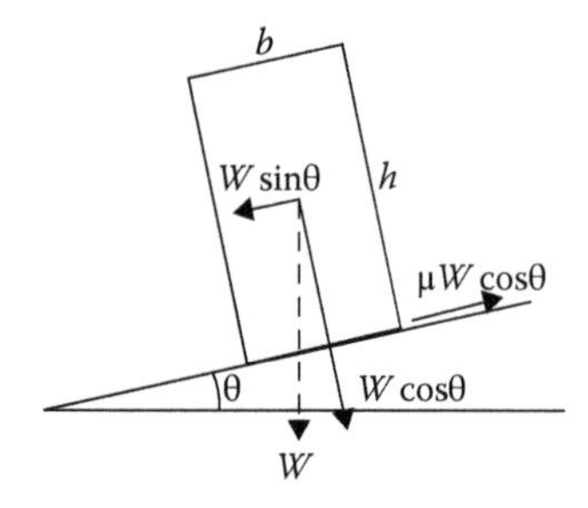

Figure 6.10 A rock block resting on a discontinuity plane.

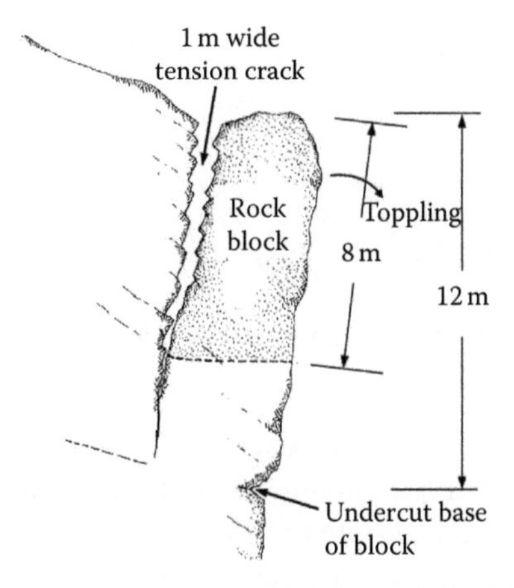

Figure 6.11 Single toppling block. (Adapted from Wyllie, D.C., *Rock Mech.*, 13, 89–98, 1980.)

Inequalities 6.27 and 6.29 define the following four conditions for toppling and/or sliding of the block:

- Toppling only: $\frac{b}{h} < \tan\theta$ and $\theta < \phi$
- Toppling with sliding: $\frac{b}{h} < \tan\theta$ and $\theta > \phi$
- Sliding only: $\frac{b}{h} > \tan\theta$ and $\theta > \phi$
- No toppling and sliding, that is stable: $\frac{b}{h} > \tan\theta$ and $\theta < \phi$

The above four conditions have also been described by Hoek and Bray (1981) and Wyllie and Mah (2004) with analysis for other types of rock toppling failure. Wyllie (1980) has presented a field situation for a single-block toppling failure (Figure 6.11).

6.4 SLOPE STABILISATION

Civil and mining engineering projects often create excavated rock slopes, which should remain stable at least up to the end of the design life of the specific project. There is a possibility during site selection to have a stable excavated rock slope without any major treatment/stabilisation, provided that the orientation of joint/bedding planes is properly assessed before site selection. Figure 6.12 illustrates how site selection for a highway project can result in stable and unstable slopes. The excavated slope should be created on the hillside only where rock strata dip away from the excavation.

Figure 6.12 Rock slopes made through excavations for highways.

There are many field situations where site selection cannot be done in view of the technical requirements/suitability; in such cases, stabilisation/treatment techniques are adopted to increase the stability of slopes. Several methods are available to increase the stability of slopes (Broms and Wong, 1991; Abramson et al., 2002; Shukla et al., 2012); these methods can be adopted singly or in combination. The choice depends primarily on the cost and the consequence of slope failure. The more commonly used slope stabilisation methods can be classified as follows (Broms and Wong, 1991):

- Geometric methods, in which the geometry of the slope is changed
- Hydrological methods, in which the groundwater table is lowered or the water content of the soil/rock is reduced
- Chemical and mechanical methods, in which the shear strength of the sliding soil/rock mass is increased or the external force causing the slope failure is reduced

Geometrical methods include slope flattening, removal of part of the soil/rock or load from the top of the slope, construction of pressure berms at the toe, terracing, replacement of slipped material by free draining material and recompaction of slip debris. Hydrological methods include the installation of surface and subsurface drains, inverted filters and thermal methods (ground freezing and heating methods). Chemical and mechanical methods include grouting, construction of restraining structures (such as concrete gravity or cantilever walls), gabion structures, crib walls, embankment piles, lime and cement columns, ground anchors, rock bolting, soil nailing and root piles, earth reinforcement and plantation of grasses and shrubs.

Wyllie and Mah (2004) classified the methods of rock slope stabilisation into the following three groups:

- Reinforcement (rock bolts, dowels, tied-back walls, shotcrete, buttresses etc.)
- Rock removal (resloping, trimming etc.)
- Protection (ditches, mesh, catch fences, warning fences, rock sheds, tunnels etc.)

Rock reinforcement is a method of adding strength to the rock in order to prevent failure. The most useful forms of reinforcement are rock bolts and anchors, which can be used on both natural and engineered slopes to prevent blocks of rock from falling/sliding away from the main mass when isolated by discontinuity planes. They are installed in such a way that the axial load in the bolt/anchor increases the effective stress at depth in soil and rock, thus improving the strength, and a component of the anchor force may also act to reduce destabilising forces and moments. In the case of fractured rock slope, rock bolts and anchors are also used in combination with reinforced concrete walls, which cover the areas of fractured rock. In Figure 6.4, the anchor force T acts to restrain the sliding rock block $A_1A_2A_3A_4$. The parametric study conducted using Equation 6.14 shows that the factor of safety (FS) of the rock slope against plane failure increases nonlinearly with an increase in T and the rate of increase is higher for lower values of seismic coefficients. It is also found that the factor of safety increases nonlinearly with an increase in its inclination (α) to the normal at the failure plane, and it becomes maximum for $\alpha \approx 70°$, beyond which it decreases.

Steel rods, known as dowels, are sometimes installed and grouted into the jointed rock to act as reinforcement. Dowels are not stressed during the installation process unlike rock bolts and anchors.

Shotcrete is a pneumatically applied, fine aggregate mortar that is usually placed in a 50–100-mm layer, and is often reinforced for improved tensile and shear strength. It is generally applied along with drain holes as a surface protection layer to the excavated rock slope face before its significant deformation or alteration, in order to provide a high strength, vary rapidly. The shotcrete effectively controls the fall of loose small rock blocks, but it provides little support against basic modes of slope failure discussed in the previous sections.

Grouting is a technique of injecting a fluid grout into the rock mass to replace all air or water present in its fissures and cracks. The grout consists of a mixture of cement and water. Sand, clay, rock flour, fly ash and other similar materials can be used as fillers in order to reduce the cost of the stabilisation work, especially where fissures and cracks are large in volume. Grouting is used ahead of development ends in mine slopes and roofs, and also in advance of tunnels driven through weak ground. If a cavity is present in the slope face, a concrete buttress can be constructed in the cavity to prevent rock falls and support the overhang, if any.

Rock slope stabilisation is often undertaken on a fairly ad hoc basis as the condition of rock mass is exposed. Some common stabilisation techniques used in practice are listed below (Fookes and Sweeney, 1976; Bromhead, 1992; Wylie and Mah, 2004):

- Flattening of overburden slope
- Trimming of unstable rock blocks
- Scaling of small loose materials/blocks

- Construction of drains and drain holes
- Use of dowels
- Installation of rock anchor to prevent sliding along discontinuity plane
- Rock bolting to strengthen the jointed rock mass
- Construction of concrete or masonry walls with weep-holes
- Construction of rock trap ditches at the toe of the slopes
- Providing rock catch fences/walls along the slope to make the adjacent areas safe for public use
- Hanging nets or chains to slow blocks tumbling
- Free hanging mesh net to guide loose rock pieces to fall down near the slope toe only
- Construction of berms/benches as a rock fall collector
- Mesh secured by bolts and gunited to protect friable formation
- Construction of rock fall barriers (gabions and concrete blocks, reinforced soil barriers etc.) at the toe of slopes
- Construction of rock sheds and tunnels
- Providing warning signals in rock fall areas

Stabilisation measures such as rock bolts and anchors prevent the detachment of rock blocks from their original position; therefore, they are classified as active measures. Walls, ditches, catch fences, rock sheds, tunnels and so on are passive measures as they do not directly interfere in the process of rock detachment, but control the dynamic effects of moving/falling blocks.

The selection of a stabilisation technique or a combination of techniques requires consideration of geotechnical (geology, rock properties, groundwater and stability analysis), construction (type of equipment, construction access, construction cost etc.) and environmental (waste disposal, aesthetics etc.) aspects. The selection is also greatly controlled by the level of stabilisation and its design life, and finally the cost. If the stabilisation work has to be effective for a longer period, the initial cost of stabilisation may be higher.

6.5 SUMMARY

1. Rock slopes can be natural or engineered (excavated). Plane, wedge, circular and toppling failures are the four basic modes of their failure.
2. Plane failures take place in rocks with well-defined discontinuities and jointing.
3. Wedge failures occur in blocky rocks with at least two continuous and nonparallel joint sets.

4. Circular failures are observed in the slopes of heavily jointed and weathered rock masses.
5. Toppling failures are generally noticed in hard rocks with regular, parallel joints dipping into the hillsides.
6. The objective of a slope stability analysis of a rock slope is to identify the most likely mode/mechanism of slope failure and to determine the associated minimum factor of safety.
7. The factor of safety of a rock slope is defined as a ratio of total force available to resist sliding of the rock block to the total force tending to induce sliding. In practice, rock slopes with FS = 1.3 to 1.5 are considered to be stable.
8. The analytical expressions for the factor of safety presented here can be used to investigate the effect of individual parameters on the stability of slopes against different modes of failure.
9. Several stabilisation/treatment measures are available for increasing the stability of slopes, and their selection depends upon several considerations, such as geotechnical, construction, environmental, level of stabilisation and cost.
10. Some examples of stabilisation measures are slope flattening, trimming and scaling, reinforcement (rock bolting and anchoring), grouting, masonry walls and gabions, drains, hanging nets and chains, rock trap ditches, rock catch fences and rock sheds and tunnels.

Review Exercises

Select the most appropriate answers to the following 10 multiple-choice questions.

1. Circular rock slope failure takes place in
   a. hard or soft rocks with well-defined discontinuities
   b. blocky rocks with at least two continuous and parallel joint sets
   c. heavily jointed and weathered rock masses
   d. hard rocks with regular, parallel joints dipping into the hillside
2. For plane rock slope failure,
   a. $i > \phi$
   b. $\theta > \phi$
   c. $i > \theta$
   d. all of the above

   where i, θ and ϕ have their usual meanings.
3. The factor of safety considered for temporary slope designs is generally
   a. 1
   b. 1.3
   c. 1.5
   d. 2

4. The factor of safety of a rock slope against plane failure does not depend on the
   a. length of the slope
   b. height of the slope
   c. inclination of the slope
   d. unit weight of the rock mass
5. Which of the following is generally considered in design practice?
   a. $k_h = k_v$
   b. $k_h = 0.5k_v$
   c. $k_h = 2k_v$
   d. $k_h < k_v$
6. The condition for a rock block to slide on a discontinuity plane is
   a. $\theta > \phi$
   b. $\theta < \phi$
   c. $\theta = \phi$
   d. $\theta \leq \phi$
7. In Figure 6.10, toppling failure only takes place when
   a. $\frac{b}{h} < \tan\theta$ and $\theta > \phi$
   b. $\frac{b}{h} < \tan\theta$ and $\theta < \phi$
   c. $\frac{b}{h} > \tan\theta$ and $\theta < \phi$
   d. $\frac{b}{h} > \tan\theta$ and $\theta > \phi$
8. The excavated slope should be created on the hillside where rock strata
   a. dip away from the excavation
   b. dip towards the excavation
   c. are vertical
   d. are horizontal
9. Which of the following is not a geometrical method of rock slope stabilisation?
   a. Slope flattening
   b. Replacement of slipped material by free draining material
   c. Rock bolting
   d. Construction of pressure berms at the toe
10. The selection of a stabilisation technique requires consideration of
    a. geotechnical aspects
    b. construction aspects
    c. environmental aspects
    d. all of the above
11. How does a natural rock slope differ from an engineered (excavated) rock slope?
12. Describe the effects of the following parameters on the stability of a rock slope against plane failure: strength parameters of

joint material, depth of tension crack and inclination of the joint plane.

13. Derive an expression for the factor of safety of a rock slope against plane failure for a generalised field situation.
14. For the rock slope shown in Figure 6.3, consider that the joint material is cohesionless, and the values of the angle of internal friction and the inclination of the joint plane to the vertical are 30° and 60°, respectively. Calculate the factor of safety of the rock slope against plane failure.
15. Consider the rock slope shown in Figure 6.4 with the following details:

    Height of the rock slope, $H = 10$ m
    Unit weight of rock, $\gamma = 20$ kN/m^3
    Surcharge pressure, $q = 100$ kN/m^2
    Stabilising force, $T = 100$ kN/m
    Depth of the tension crack, $z = 2.5$ m
    Depth of water in the tension crack, $z_w = 2.5$ m
    Angle of inclination of stabilising force to the normal at the failure plane, $\alpha = 40°$
    Angle of shearing resistance of the joint material, $\phi = 25°$
    Cohesion of the joint plane material, $c = 32$ kN/m^2
    Angle of inclination of the slope face to the horizontal, $i = 50°$
    Angle of inclination of the joint plane/failure plane to the horizontal, $\theta = 35°$
    Horizontal seismic coefficient, $k_h = 0.2$
    Vertical seismic coefficient, $k_v = 0.1$

    Calculate the factor of safety of the rock slope against plane failure. Assume that the height of tension crack is one-fourth of the height of the rock slope, and the tension crack is completely filled with water.
16. Discuss about the optimum inclination of the anchor used for stabilising a sliding rock block, separated by a sloping joint/bedding plane.
17. What is the difference between plane and wedge failures? Which one is the most common failure in field?
18. Derive an expression for the factor of safety of a rock slope against a simple wedge failure.
19. For the rock slope shown in Figure 6.7, consider the following: $i = 60°$, $\beta = 40°$, $\phi = 38°$, $\theta_A = 40°$ and $\theta_B = 45°$. Calculate the factor of safety of the rock slope against wedge failure.
20. Is there any difference between slope failures in soils and rocks? Explain.
21. What are the different types of toppling failures? Explain with the help of neat sketches.
22. Discuss the conditions for toppling and sliding of a rock block resting in a joint plane.
23. What are the different rock slope stabilisation techniques and their classifications?

24. Discuss the suitability of rock bolting and anchoring for stabilising the rock slopes.
25. What is the difference between a dowel and a rock bolt?
26. What is shotcrete? How does it differ from grouting?
27. Enumerate the factors that are considered for the selection of a stabilisation technique for a specific field application.

Answers:
1. c; 2. d; 3. b; 4. a; 5. c; 6. a; 7. b; 8. a; 9. c; 10. d
14. 1
15. 1.17
19. 126

REFERENCES

Abramson, L.W., Lee, T.S., Sharma, S. and Boyce, G.M. (2002). *Slope Stability and Stabilization Methods*. John Wiley and Sons, Inc., New York.

Alejano, L.R., Gomez-Marquez, I. and Martinez-Alegria, R. (2010). Analysis of a complex toppling-circular failure. *Engineering Geology*, Vol. 114, pp. 93–104.

Bathurst, R.J., Hatami, K. and Alfaro, M.C. (2012). Geosynthetics-reinforced soil walls and slopes—seismic aspects, in *Handbook of Geosynthetic Engineering*, 2nd edition, Shukla, S. K., editor, ICE Publishing, London, pp. 317–363.

Bromhead, E.N. (1992). *The Stability of Slopes*, 2nd edition, Blackie Academic & Professional, Glasgow.

Broms, B.B. and Wong, K.S. (1991). Landslides, in *Foundation Engineering Handbook*, Fang, H.Y., editor, Van Nostrand Reinhold, New York, pp. 410–446.

Das, B.M. (2012). *Fundamentals of Geotechnical Engineering*, 4th edition, Cengage Learning, Stamford.

Fookes, P.G. and Sweeney, M. (1976). Stabilization and control of local rockfalls and degrading rock slopes. *Quarterly Journal of Engineering Geology*, Vol. 9, pp. 37–56.

Goodman, R.E. (1989). *Introduction to Rock Mechanics*, 2nd edition, Wiley, New York.

Goodman, R.E. and Kieffer, D.S. (2000). Behaviour of rock in slopes. *Journal of Geotechnical and Geoenvironmental Engineering*, Vol. 126, No. 8, pp. 675–684.

Hoek, E. and Bray, J. (1981). *Rock Slope Engineering*, 3rd edition, Taylor & Francis, London.

Hossain, M.M. and Shukla, S.K. (2010). Effect of vertical seismic coefficient on the stability of rock slopes against plane failure. *Proceedings of the 6th Australasian Congress on Applied Mechanics, ACAM 6*, December 12–15, 2010, Perth Convention Centre, Perth, Western Australia, Paper No. 1108.

Lambe, T.W. and Whitman, R.V. (1979). *Soil Mechanics*, SI version, John Wiley & Sons, New York.

Ling, H.I. and Cheng, A.H.D. (1997). Rock sliding induced by seismic force. *International Journal of Rock Mechanics and Mining Sciences*, Vol. 34, No. 6, pp. 1021–1029.

Shukla, S.K. and Hossain, M.M. (2010). Design charts for rock slope stability against plane failure under seismic loading condition. *Proceedings of the ISRM International Symposium 2010 and 6th Asian Rock Mechanics Symposium*, October 23–27, 2010, New Delhi, India, Paper No. 64.

Shukla, S.K. and Hossain, M.M. (2011a). Analytical expression for factor of safety of an anchored rock slope against plane failure. *International Journal of Geotechnical Engineering*, Vol. 5, No. 2, pp. 181–187.

Shukla, S.K. and Hossain, M.M. (2011b). Stability analysis of multi-directional anchored rock slope subjected to surcharge and seismic loads. *Soil Dynamics and Earthquake Engineering*, Vol. 31, Nos. 5–6, pp. 841–844.

Shukla, S.K., Khandelwal, S., Verma, V.N. and Sivakugan, N. (2009). Effect of surcharge on the stability of anchored rock slope with water filled tension crack under seismic loading condition. *Geotechnical and Geological Engineering, an International Journal*, Vol. 27, No. 4, pp. 529–538.

Shukla, S.K., Sivakugan, N. and Das, B.M. (2012). Slopes–stabilization, in *Handbook of Geosynthetic Engineering*, 2nd edition, Shukla, S.K., editor, ICE Publishing, London, pp. 223–243.

Taylor, D.W. (1948). *Fundamentals of Soil Mechanics*. John Wiley & Sons, New York.

Terzaghi, K. (1943). *Theoretical Soil Mechanics*. John Wiley & Sons, New York.

Terzaghi, K., Peck, R.B. and Mesri, G. (1996). *Soil Mechanics in Engineering Practice*. John Wiley & Sons, New York.

Wyllie, D.C. (1980). Toppling rock slope failures: Examples of analysis and stabilization. *Rock Mechanics*, Vol. 13, pp. 89–98.

Wyllie, D.C. and Mah, C.W. (2004). *Rock Slope Engineering*, 4th edition, Spon Press, London.

Chapter 7

Foundations on rock

7.1 INTRODUCTION

The word 'foundation' refers to the load-carrying structural member of an engineering system (e.g., building, bridge, road, runway, dam, pipeline, tower or machine) below the ground surface as well as the earth mass that finally supports the loads of the engineering system.

In Chapter 1, it is explained that rock is a hard, compact, naturally occurring earth material composed of one or more minerals and is permanent and durable for engineering applications. Most rocks generally require blasting for their excavation. In general, a site consisting of rocks is usually recognised as the best foundation site for supporting structures because of the ability of rocks to withstand much higher loads than the soils. We described in Chapters 1 and 4 that in situ rocks carry different types of discontinuities and planes of weakness (Figure 7.1) such as joints, fractures, bedding planes and faults and therefore, they are often nonhomogeneous and anisotropic in their in situ properties at construction sites. This is the reason why it has not been possible to analyse foundations on rock in a generalised form. This chapter presents the description of shallow and deep foundations on rock to explain their fundamentals and some commonly used approaches to estimate the design value of their load-carrying capacity.

7.2 SHALLOW FOUNDATIONS

7.2.1 Meaning of shallow foundation

A foundation is considered shallow if its depth (D) is generally less than or equal to its width (B). Therefore, for a shallow foundation,

$$\frac{D}{B} \leq 1 \tag{7.1}$$

(a) (b)

Figure 7.1 Rock foundation for a barrel aqueduct at 44.900 km of the Bansagar Feeder Channel, Sidhi District, Madhya Pradesh, India: (a) foundation trench and (b) rock condition at the founding level. (After Shukla, S.K., Allowable Load-Bearing Pressure for the Foundation of Barrel Aqueduct on Rock at km 44.900 of the Bansagar Feeder Channel, Dist. Sidhi, MP, India. A technical report dated 29 June 2007, Department of Civil Engineering, Institute of Technology, Banaras Hindu University, Varanasi, India, 2007.)

In practice, the ratio D/B of a foundation can be greater than unity and still be treated as a shallow foundation. The authors consider that a foundation can be described as shallow if its depth is less than or equal to about 3.5 m below the ground surface.

7.2.2 Types of shallow foundations

The most common types of shallow foundations on rock and soil are *spread footings* and *mats* (or *rafts*). A *spread footing* is simply an enlargement of a load-bearing wall or column that makes it possible to spread the load of the engineering system or structure over a large area of the rock and soil. The spread footing for supporting a long wall is called *strip footing*, which may have a length-to-width ratio more than 5. A *mat* or *raft foundation* is a continuous slab constructed over the rock or soil bed to support an arrangement of columns and walls in a row or rows (Figure 7.2). Mat foundations are preferred for weak soils and heavily jointed and fractured rock masses that have low bearing capacities but that will have to support high column and/or wall loads. A mat that supports two columns is called *combined footing*. Mat foundations undergo significantly reduced differential settlements compared to those for spread footings.

7.2.3 Depth of foundation

For shallow foundations resting on a rock, the depth of the rock, which is weathered or fissured, is generally excluded in deciding the depth of

Figure 7.2 A raft foundation for an aqueduct under construction at 46.615 km of the Bansagar Feeder Channel, Sidhi District, Madhya Pradesh, India. (After Shukla, S.K., Allowable Load-Bearing Pressure for the Foundation of Aqueduct on Rock/Soil at km 46.615 of the Bansagar Feeder Channel, Dist. Sidhi, MP, India. A technical report dated 7 June 2006, Department of Civil Engineering, Institute of Technology, Banaras Hindu University, Varanasi, India, 2006.)

foundation in the rock. The foundation level is established at sufficient depth so as to ensure that they do not get undermined, keeping in view the continued erosion of the rock bed. In hard rocks, with ultimate compressive strength of 10 MPa or above arrived at after considering the overall characteristics of the rock, such as fissures, joints and bedding planes, the minimum depth of foundation is taken as 0.6 m, whereas in all other types of rock, it is 1.5 m.

7.2.4 Load-bearing capacity terms

The load per unit area at the base level of foundation that causes shear failure to occur in the earth mass (soil or rock) is termed the *ultimate bearing capacity* (q_u) of the foundation. This capacity depends on the characteristics of the earth mass and is also governed by the geometric dimensions of the foundation and its depth below the ground surface. The *safe bearing capacity* (q_s) is the pressure at the base level of foundation that can be safely carried by the foundation without shear failure of the earth mass. The load per unit area at the base level of foundation that causes permissible or specified settlement of the engineering system is called the *safe bearing pressure* (q_p). The lower of the safe bearing capacity and the safe bearing pressure is called the *allowable bearing pressure* (q_a). If the ultimate bearing capacity, safe bearing capacity, safe bearing pressure and allowable bearing pressure are estimated by deducting the *effective overburden pressure* at the base level of foundation, they are termed the *net ultimate bearing capacity* (q_{nu}), *net safe bearing capacity*

(q_{ns}), *net safe bearing pressure* (q_{np}) and *net allowable bearing pressure* (q_{na}), respectively. The value of the net allowable bearing pressure (q_{na}) is generally recommended for design of shallow foundations.

7.2.5 Estimation of load-bearing capacity

The compressive strength of rocks ranges from less than 10 MPa to more than 300 MPa (see Figure 3.7 of Chapter 3). If a construction site consists of strong/hard rock, shallow foundations such as spread footings can support substantial loads; however, the presence of a single discontinuity plane in a particular direction (Figure 7.1) can cause sliding failure of the foundation. Discontinuities in rock also causes reduced bearing capacity of the foundation supported by the rock. Rock without discontinuities rarely occurs at or near the ground surface at the specific construction site. Therefore, it becomes essential to estimate the realistic values of the bearing capacity of foundations on rock, considering the presence of discontinuities.

The bearing capacity of foundations on rock consisting of weaknesses is difficult to determine because of wide variations in the weaknesses from site to site and from location to location within a site resulting from nonhomogeneity and anisotropic characteristics. Usually, the net allowable bearing pressure to be used for design is restricted by the *local building code*; however, geology, rock type and quality (as RQD), as discussed in Chapters 1 and 3, are significant parameters that should be used together with the recommended code value. With the exception of a few porous limestone and volcanic rocks and some shales, the strength of bedrock in situ is greater than the compressive strength of the foundation concrete. Therefore, design values of net allowable bearing pressure are often limited by the strength of concrete. If concrete foundation is submerged under water, the bearing value of concrete should be reduced, and the allowable bearing pressure of foundation on rock is further complicated by the possibility of rock softening. The common sandstones and limestones have modulus of elasticity values from that of a poor concrete to high strength concrete. Very hard igneous and metamorphic rocks exhibit considerably greater modulus of elasticity values. Almost all rocks can withstand a compressive stress higher than concrete; the following are some of exceptions (Teng, 1962):

1. Limestones with cavities and fissures, which may be filled with clay or silt.
2. Rocks with bedding planes, folds, faults or joints at an angle with the bottom of the footing.
3. Soft rocks that reduce their strength after wetting; weathered rocks, which are very treacherous, and shale, which may become clay or silt in a matter of hours of soaking.

Some attempts have been made to present the theoretical solution for the bearing capacity of strip footings on jointed rock masses (Yu and Sloan, 1994; Prakoso and Kulhawy, 2004). The theoretical approach requires idealisation of the strength of the intact rock and strength, spacing and orientation of the discontinuities. Because of a wide variation of these factors, it is rarely possible to present a generalised bearing capacity equation for foundations on rock in the way it is done for foundations on soil. In practice, empirical approaches of estimating the allowable bearing pressure are widely used, and some of them are discussed here.

Settlement of rock foundation is more often of concern than its bearing capacity. Therefore, for shallow foundations on rock, it is generally found that $q_{np} < q_{ns}$, therefore, $q_{na} = q_{np}$. If q_{np} is calculated based on the plate load test (Shukla and Sivakugan, 2011), the permissible settlement is taken as 12 mm even for larger loaded areas (BIS, 2005). In the case of rigid structures such as reinforced concrete silos, the permissible settlement may be increased judiciously, if required. If the spacing of discontinuities in rock foundation is wide (1–3 m) or very wide (>3 m), q_{np} for preliminary design of shallow foundations on rock can be determined from the classification of rock mass as given in Table 7.1.

The Indian Roads Congress suggests that the allowable pressure values of rocks for average condition may be taken as follows (IRC, 2000): for hard rocks, $q_{na} = 2–3$ MPa; for soft rocks, $q_{na} = 1–2$ MPa and for weathered rocks, conglomerates, and laterites, $q_{na} < 1$ MPa. These values should be modified after taking into account the various characteristics of rocks at the construction site.

If the spacing of discontinuities in rock foundation is moderately close (0.3–3 m), q_{np} for design of shallow foundations on rock can be determined from the strength of the rock cores obtained during subsurface investigation.

Table 7.1 Net safe bearing pressure based on classification

Type of Rock	*Net Safe Bearing Pressure, q_{np} (MPa)*
Massive crystalline bedrock including granite, diorite and gneiss	10
Foliated rocks such as schist and slate in sound condition	4
Bedded limestone in sound condition	4
Sedimentary rocks including hard shale and sandstone	2.5
Soft or heavily fractured bedrock (excluding shale) and soft limestone	1
Soft shale	0.4

Source: Adapted from BIS, *Code of Practice for Design and Construction of Shallow Foundations on Rocks.* IS: 12070–1987 (Reaffirmed 2005), Bureau of Indian Standards (BIS), New Delhi, India, 2005.

If $q_{u(av)}$ is the average unconfined compressive strength of rock cores, the safe bearing pressure, q_p, can be given as

$$q_p = q_{u(av)} N_d \tag{7.2}$$

where N_d is an empirical coefficient depending on the spacing of discontinuities and is expressed as

$$N_d = \frac{3 + S/B}{10\sqrt{1 + 300(\delta/S)}} \tag{7.3}$$

where δ is the thickness (aperture) of discontinuities, S is the spacing of discontinuities and B is the width of footing. For spacing of discontinuities of 0.3–1, 1–3 and 3 m, the typical values of N_d are 0.1, 0.25 and 0.4, respectively. It may be noted that Equation 7.2 is valid under the following six conditions (BIS, 2005):

1. The rock surface is parallel to the base of the foundation.
2. The structural load is normal to the base of the foundation.
3. The spacing of discontinuities is greater than 0.3 m.
4. The aperture (opening) of discontinuities is less than 10 mm (15 mm if filled with soil and rock debris).
5. The foundation width is greater than 0.3 m.
6. The factor of safety is 3.

EXAMPLE 7.1

A strip footing of 1.2 m width rests on the bedrock exposed to the ground surface. The bedrock is horizontally bedded with spacing $S = 0.8$ m, aperture $\delta = 8$ mm and $q_{u(av)} = 80$ MPa. Estimate the safe bearing pressure.

Solution

From Equation 7.3,

$$N_d = \frac{3 + (0.8)/(1.2)}{10\sqrt{1 + 300(0.008/0.8)}} = 0.09$$

From Equation 7.2,

$$q_p = (80)(0.09) = 7.2\,\text{MPa} = 7200\,\text{kN/m}^2$$ **Answer**

In many cases, the allowable bearing pressure is taken in the range of one-third to one-tenth the unconfined compressive strength obtained from

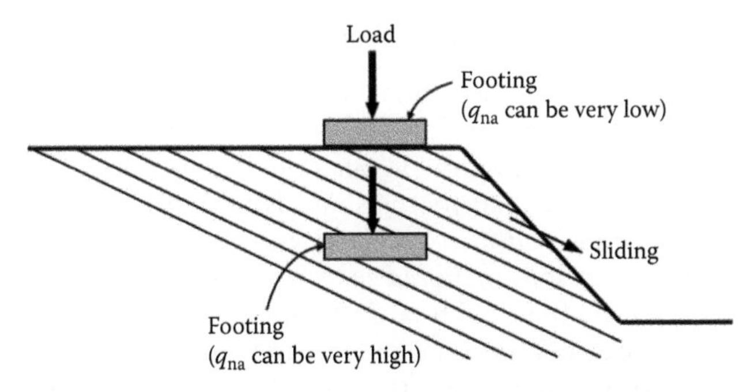

Figure 7.3 An example of the importance of consideration of geological condition and presence of discontinuities while recommending the net allowable bearing pressure for design of foundations on rock.

intact rock samples and using RQD as a guide, for example, as one-tenth for a small RQD (Bowles, 1996). When the RQD of the foundation rock tends to zero, one should treat it as soil mass and obtain the allowable bearing pressure using the bearing capacity theories for soils.

While recommending the allowable bearing pressure, it is important that the geological conditions and discontinuities present at the rock foundation site be analysed properly because they greatly control the net allowable bearing pressure compared to the strength of intact rock mass. For example, in Figure 7.3, the rock foundation consists of rock beds dipping away from the slope, and therefore, a surface footing may be unstable due to the possible slides of the underlying top rock beds, while a footing at some depth may be stable. The readers can refer to the book by Wyllie (1999) for more geological details.

7.3 DEEP FOUNDATIONS

7.3.1 Meaning of deep foundation

The foundation is considered as deep if its depth (D) is generally greater than its width (B). Therefore, for a deep foundation,

$$\frac{D}{B} > 1 \tag{7.4}$$

The authors consider that a foundation can be described as deep if its depth is greater than about 3.5 m below the ground surface.

When the soil near the ground surface is highly compressible and too weak to support the load transmitted by the superstructure, deep foundations are used to transmit the load to the underlying stronger soil layer or the bedrock.

7.3.2 Types of deep foundations

The most common types of deep foundations on rock and soil are *piles* and *drilled piers*. *Piles* are structural members that are made of steel, concrete and/or timber. Placing a structure on pile foundations is much more expensive than having it on spread footings and is likely to be more expensive than a raft foundation. A *drilled pier* (also known as a *drilled shaft*, *drilled caisson* or simply *caisson*, or *bored pile*) is a cast-in-place pile, generally having a diameter of about 2.5 ft ($\approx$ 750 mm) or more. It is constructed by drilling a cylindrical hole into the ground and subsequently filling it with concrete along with reinforcement (Figure 7.4) or no reinforcement.

If subsurface records establish the presence of rock or rock-like material at a site within a reasonable depth, piles are generally extended to the bedrock and socketed properly, if required (Figure 7.5a). In this case, based on the strength of bedrock or rock-like material, the *ultimate load-carrying capacity* (Q_u) of the piles depends entirely on the load-bearing capacity of the bedrock or rock-like material, and the piles are called *point-bearing piles or end-bearing piles*, and therefore it is given as

$$Q_u = Q_p \tag{7.5a}$$

(a)

(b)

Figure 7.4 A bored pile/drilled pier in fractured and weathered rock under construction at 52.106 km of the Bansagar Feeder Channel, Sidhi District, Madhya Pradesh, India: (a) before concrete filling and (b) after concrete filling. (After Shukla, S.K., The pile termination at km 52.106 of the Bansagar Feeder Channel, Dist.–Sidhi, MP, India. A technical report dated 17 December 2008, Department of Civil Engineering, Institute of Technology, Banaras Hindu University, Varanasi, India, 2008.)

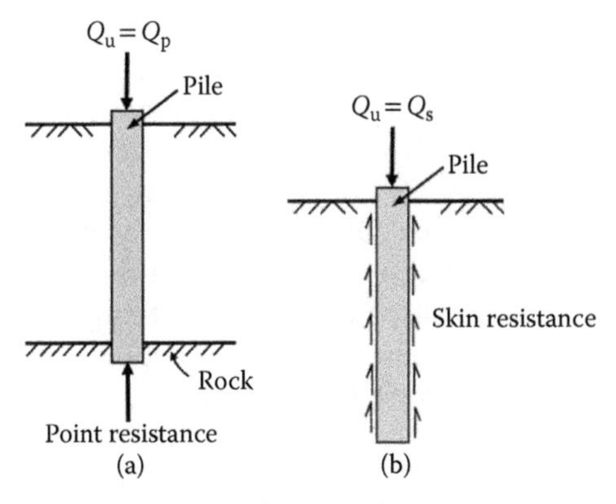

Figure 7.5 (a) Point-bearing pile and (b) friction pile.

where Q_p is the *load-carrying capacity of the pile point/tip*, that is, the *point capacity* or *end-bearing capacity* of the pile.

When bedrock or rock-like material is not available at a reasonable depth below the ground surface, piles can be designed to transmit the structural load through friction and/or adhesion to the soil adjacent to the pile only or to both the adjacent soil and the underlying firm soil stratum, if available. The piles that transmit loads to the adjacent soil through friction and/or adhesion are called *friction piles* (Figure 7.5b) and therefore

$$Q_u = Q_s \tag{7.5b}$$

where Q_s is frictional resistance of the pile.

The piles in heavily jointed/fractured and weathered rocks where bedrock does not exist at a reasonable depth are generally designed considering them as both point-bearing and friction piles, the way they are designed in soils; thus, the *ultimate load-carrying capacity* of the pile is given as

$$Q_u = Q_p + Q_s \tag{7.6}$$

The estimation of Q_p and Q_s for piles in soils including heavily jointed/fractured and weathered rocks that behave similar to soils is described in detail in most geotechnical books. This chapter discusses the estimation of load-carrying capacity of piles resting on bedrock only.

7.3.3 Estimation of load-carrying capacity

A pile resting on bedrock or rock-like material is generally designed to transfer large structural loads, and its ultimate load-carrying capacity is calculated as only the point capacity or end-point capacity Q_p (Equation 7.5a). In general, the point capacity of a pile resting on bedrock or rock-like material is calculated in the following two steps: (1) capacity based on strength of rock or rock-like material and (2) capacity based on the yield strength of the pile material; the lower value is taken as the design value of point capacity. Unless a pile is bearing on soft rock such as shale or other poor quality rocks (RQD < 50), the capacity calculated from the strength of rock is higher than that calculated from the yield strength of the pile material. Therefore, in most cases, calculation of the load-carrying capacity of the pile resting on rock based on the yield strength of the pile material is sufficient (Kumar, 2011). The ultimate unit point resistance in rock is approximately (Goodman, 1980; Das, 2013)

$$q_p = q_u(N_\phi + 1) \tag{7.7}$$

where q_u is the unconfined compression strength of rock,

$$N_\phi = \tan^2(45° + \phi/2) \tag{7.8}$$

and ϕ is the drained angle of internal friction.

The unconfined compressive strength of rock is generally determined in the laboratory by conducting unconfined compression strength tests on small diameter cylindrical intact rock specimens prepared from rock samples collected during subsurface investigation. It is observed that the unconfined compressive strength of rock decreases as the diameter of laboratory rock specimen increases, which is referred to as the *scale effect*. For rock specimens larger than about 1 m in diameter, the value of q_u remains approximately constant. There appears to be a fourfold to fivefold reduction in the magnitude of q_u in this process. The scale effect is primarily caused by randomly distributed large and small fractures and also by progressive ruptures along the slip lines. Hence, it is generally recommended that

$$q_{u(design)} = \frac{q_{u(lab)}}{5} \tag{7.9}$$

Table 7.2 lists $q_{u(lab)}$ and ϕ values for some rocks. Substituting q_u in Equation 7.7 by $q_{u(design)}$ from Equation 7.9,

$$q_p = \left[\frac{q_{u(lab)}}{5}\right](N_\phi + 1) \tag{7.10}$$

Table 7.2 Typical values of laboratory unconfined compressive strength and drained friction angle of some rocks

Rock Type	*Unconfined Compressive Strength, q_u (MPa)*	*Drained Angle of Internal Friction ϕ (Degrees)*
Sandstone	70–140	27–45
Limestone	105–210	30–40
Shale	35–70	10–20
Granite	140–210	40–50
Marble	60–70	25–30

The point capacity or end-bearing capacity of the pile is

$$Q_p = q_p A_p \tag{7.11}$$

where A_p is the area of the pile point. Substituting q_p from Equation 7.10 in Equation 7.11,

$$Q_p = \left[\frac{q_{u(lab)}}{5}\right](N_\phi + 1)A_p \tag{7.12}$$

From Equations 7.5 and 7.12,

$$Q_u = \left[\frac{q_{u(lab)}}{5}\right](N_\phi + 1)A_p \tag{7.13}$$

The design load-carrying capacity or allowable load-carrying capacity of a pile is defined as

$$Q_a = \frac{Q_u}{FS} \tag{7.14}$$

where FS is a factor of safety, depending on the uncertainties of estimation of Q_u. It is common to use large safety factors in estimating the load-carrying capacity of rock foundation. The FS should be somewhat dependent on RQD, defined in Chapters 1 and 3. For example, an RQD of 80% would not require as high an FS as for RQD = 40%. It is common to use FS from 2.5 to 10.

From Equations 7.13 and 7.14,

$$Q_a = \left[\frac{q_{u(lab)}}{5}\right]\left[\frac{(N_\phi + 1)A_p}{FS}\right] \tag{7.15}$$

Based on the yield strength (f_y) of the pile material, the ultimate load-carrying capacity of the pile is given as

$$Q_u = f_y A_p \tag{7.16}$$

From Equations 7.14 and 7.16,

$$Q_a = \frac{f_y A_p}{FS} \tag{7.17}$$

The values of Q_a calculated from Equations 7.15 and 7.17 are compared, and the lower value is taken as the allowable point capacity of the pile for its design.

EXAMPLE 7.2

A pile of diameter of 60 cm and length of 10 m passes through the highly jointed and weathered rock mass and rests on a shale bed. For shale, laboratory unconfined compressive strength = 38 MPa and drained friction angle = 26°. Estimate the allowable point capacity of the pile. Assume that the pile material has sufficient strength and use a factor of safety of 5.

Solution

Given that diameter D = 60 cm = 0.6 m, length L = 10 m, $q_{u(lab)}$ = 38 MPa and ϕ = 13°, the area of the pile tip is

$$A_p = \left(\frac{\pi}{4}\right)D^2 = \left(\frac{3.14}{4}\right)(0.6)^2 = 0.2826 \text{ m}^2$$

From Equation 7.8,

$$N_\phi = \tan^2(45° + 13°) = 2.56$$

From Equation 7.15,

$$Q_a = \left[\frac{38}{5}\right]\left[\frac{(2.56 + 1)(0.2826)}{5}\right] = 1.529\,\text{MN} = 1529\,\text{kN}$$ **Answer**

7.4 FOUNDATION CONSTRUCTION AND TREATMENT

The excavation of rocks for the foundation trench requires that they should be fragmented first by drilling and loading or by controlled blasting without any damage to adjacent structures, if any. The excavation procedure is highly governed by the geological features of the site, as explained in Chapter 1, and by the experience of the person doing the excavation work.

Vertical (open or soil-filled) joints are commonly present even in unweathered rocks. Such joints beneath the shallow foundations should be cleaned out to a depth of four to five times their width and filled with *slush grout* (cement–sand mixture in 1:1 ratio by volume with enough water). Grouting is also usually carried out where the shallow foundation bears on rock containing voids to strengthen the rock. Larger spaces, wider at the top, are likely to occur at intersecting joints, which are commonly filled with *dental concrete* (stiff mixture of lean concrete) placed and shaped by shovel. If horizontal joints are located beneath the shallow foundation, such joints may lead to differential and sudden settlements. If the estimated settlement exceeds the permissible limit, the rock above the joints may be removed provided this task is economical; otherwise, deep foundations may be recommended.

If bedded limestones are present at the foundation site, there might be a possibility of solution cavities, which require a detailed investigation. Such cavities may be filled with cement grout. Solution cavities may render the foundation trench bed uneven; in that situation, the depth of foundation should be taken to a level such that at least 80% rock area is available to support the foundation. It is important to ensure that the base of the foundation does not overhang at any corner. If the filled-up soil and loose pockets of talus deposit are present at the foundation site, they should be excavated, cleaned and backfilled with lean concrete of required strength. If a foundation has to rest on a sloping rock, special attention should be paid to the discussion of the stability of slopes in Chapter 6.

For more geotechnical aspects of foundations on rock, refer to *Foundation Engineering* by Peck et al. (1974).

7.5 SUMMARY

1. A foundation is considered shallow if its depth is generally less than or equal to its width. The most common types of shallow foundations on rock and soil are *spread footings* and *mats* (or *rafts*).
2. In hard rocks, with ultimate compressive strength of 10 MPa or above arrived at after considering the overall characteristics of the rock, such as fissures, joints and bedding planes, the minimum depth of foundation is taken as 0.6 m, whereas in all other types of rock, it is 1.5 m.

3. The value of net allowable bearing pressure (q_{na}) is generally recommended for design of shallow foundations. The allowable pressure values of rocks for average conditions may be taken as follows: for hard rocks, q_{na} = 2–3 MPa; for soft rocks, q_{na} = 1–2 MPa and for weathered rocks, conglomerates and laterites, q_{na} < 1 MPa. These values should be modified after taking into account the various characteristics of rocks at the construction site.
4. In many cases, the allowable bearing pressure is taken in the range of one-third to one-tenth the unconfined compressive strength obtained from intact rock samples and using RQD as a guide, for example, as one-tenth for a small RQD.
5. The foundation is considered deep if its depth is generally greater than its width. The most common types of deep foundations on rock and soil are *piles* and *drilled piers.*
6. In most cases, calculation of the load-carrying capacity of the pile resting on rock based on the yield strength of the pile material is sufficient.
7. It is common to use large safety factors (2.5–10) in estimating the bearing capacity of rock foundation.
8. The foundation excavation and treatment procedure is highly governed by the geological features of the site as well as by the experience of the person doing the excavation work.

Review Exercises

Select the most appropriate answers to the following 10 multiple-choice questions.

1. Which of the following ratios of width to depth of a foundation does not refer to a shallow foundation?
   a. 0.5
   b. 1.0
   c. 2.0
   d. Both (b) and (c)
2. A high rise building site consists of heavily jointed and fractured rock mass. The most suitable foundation for this site will be
   a. strip footing
   b. isolated square/rectangular footing
   c. raft foundation
   d. all of the above
3. Core drilling was carried out at a rock foundation site, and the RQD was estimated to be 25%. What will be the minimum depth of foundation at this site?
   a. 0.6 m
   b. 0.75 m
   c. 1 m
   d. 1.5 m

4. For the design of shallow foundation, which of the following value is generally recommended?
   a. Safe bearing capacity
   b. Net allowable bearing pressure
   c. Allowable bearing pressure
   d. Safe bearing pressure
5. The net safe bearing pressure of bedded limestone bedrock is generally
   a. 0.4 MPa
   b. 1 MPa
   c. 2.5 MPa
   d. 4 MPa
6. A drilled pier is also known as a
   a. drilled shaft
   b. drilled caisson
   c. caisson
   d. all of the above
7. For a point-bearing pile, the ratio of ultimate load-carrying capacity to the point capacity is
   a. equal to 0.5
   b. equal to 1
   c. less than 1
   d. greater than 1
8. The drained angle of friction (in degrees) for limestone ranges from
   a. 10 to 20
   b. 20 to 30
   c. 30 to 40
   d. 40 to 50
9. The factor of safety used in estimating the bearing capacity of rock foundation ranges from
   a. 1 to 2
   b. 2 to 4
   c. 2.5 to 10
   d. None of the above
10. Vertical joints in rock foundations are generally filled with slush grout that has cement–sand mixture in the volume ratio of
    a. 1:1
    b. 1:1.5
    c. 1.2
    d. 1:3
11. What is meant by the term 'foundation'? Explain briefly.
12. Differentiate between shallow and deep foundations.
13. What type of shallow foundation would you recommend for a building on a heavily jointed and fractured rock site?
14. What should be the minimum depth of foundation on hard bedrock?
15. Define the following terms: ultimate bearing capacity, safe bearing capacity, safe bearing pressure and allowable bearing pressure.

16. Define the following terms: net ultimate bearing capacity, net safe bearing capacity, net safe bearing pressure and net allowable bearing pressure.
17. What are the parameters that govern the bearing capacity of foundations on rock?
18. A strip footing of 1.5 m width rests on bedrock exposed to the ground surface. The bedrock is horizontally bedded with spacing $S = 1$ m, aperture $\delta = 10$ mm and $q_{u(av)} = 60$ MPa. Estimate the safe bearing pressure.
19. How do geological site conditions affect the bearing capacity of rock foundation? Explain giving some field examples.
20. How does a point-bearing pile differ from a friction pile? Explain with the help of neat sketches.
21. Explain the method of estimating the point-bearing capacity of a pile resting on rock.
22. A pile of diameter of 50 cm and length of 12 m passes through the highly jointed and weathered rock mass and rests on a sandstone bed. For sandstone, laboratory unconfined compressive strength = 90 MPa and drained friction angle = 38°. Estimate the allowable point capacity of the pile. Assume that the pile material has sufficient strength and use a factor of safety of 5.
23. Is it possible to excavate rock without blasting? Can you suggest some methods?
24. How are vertical joints in rock foundation treated before the construction of structural footings?
25. How will you deal with solution cavities located at a limestone foundation site?

Answers:
1. a; 2. c; 3. d; 4. b; 5. d; 6. d; 7. b; 8. c; 9. c; 10. a
18. 10.8 MPa
22. 3675 kN

REFERENCES

BIS. (2005). *Code of Practice for Design and Construction of Shallow Foundations on Rocks.* IS: 12070–1987 (Reaffirmed 2005), Bureau of Indian Standards (BIS), New Delhi, India.

Das, B.M. (2013). *Fundamentals of Geotechnical Engineering.* 4th edition, Cengage Learning, Stamford.

Goodman, R.E. (1980). *Introduction to Rock Mechanics.* Wiley, New York.

Kumar, S. (2011). Design of Pile Foundations, in *Handbook of Geotechnical Engineering*, B.M. Das, editor, J. Ross Publishing, Inc., Fort Lauderdale, FL, pp. 5.1–5.73.

Peck, R.B., Hanson, W.E. and Thornburn, T.H. (1974). *Foundation Engineering.* 2nd edition, John Wiley & Sons, Inc., New York.

Prakoso, W.A. and Kulhawy, F.H. (2004). Bearing capacity of strip footings on jointed rock masses. *Journal of Geotechnical and Geoenvironmental Engineering*, ASCE, Vol. 130, No. 12, pp. 1347–1349.

Shukla, S.K. (2006). Allowable Load-Bearing Pressure for the Foundation of Aqueduct on Rock/Soil at km 46.615 of the Bansagar Feeder Channel, Dist. Sidhi, MP, India. A technical reported dated 7 June 2006, Department of Civil Engineering, Institute of Technology, Banaras Hindu University, Varanasi, India.

Shukla, S.K. (2007). Allowable Load-Bearing Pressure for the Foundation of Barrel Aqueduct on Rock at km 44.900 of the Bansagar Feeder Channel, Dist. Sidhi, MP, India. A technical reported dated 29 June 2007, Department of Civil Engineering, Institute of Technology, Banaras Hindu University, Varanasi, India.

Shukla, S.K. (2008). The pile termination at km 52.106 of the Bansagar Feeder Channel, Dist.–Sidhi, MP, India. A technical reported dated 17 December 2008, Department of Civil Engineering, Institute of Technology, Banaras Hindu University, Varanasi, India.

Shukla, S.K. and Sivakugan, N. (2011). Site Investigation and in situ Tests, in *Geotechnical Engineering Handbook*, B.M. Das, editor, J. Ross Publishing, Inc., FL, pp. 10.1–10.78.

Teng, W.C. (1962). *Foundation Design*. Prentice-Hall of India Pvt. Ltd., New Delhi.

Wyllie, D.C. (1999). *Foundations on Rock*. E&FN Spon, London.

Yu, H.S. and Sloan, S.W. (1994). Bearing capacity of jointed rock. *Proceeding of the 8th International Conference on Computer Methods and Advances in Geomechanics*, Balkema, Rotterdam, The Netherlands, Vol. 3, pp. 2403–2408.

Appendix A

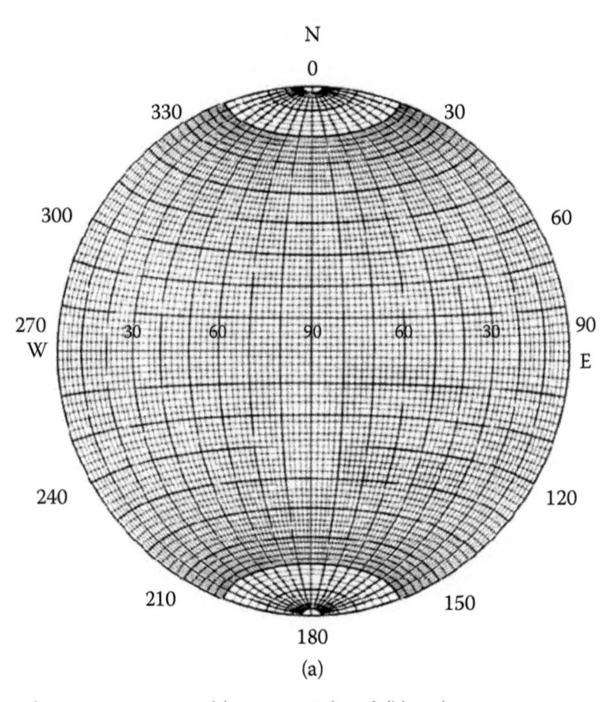

Figure A.1 Equal area stereonets: (a) equatorial and (b) polar.

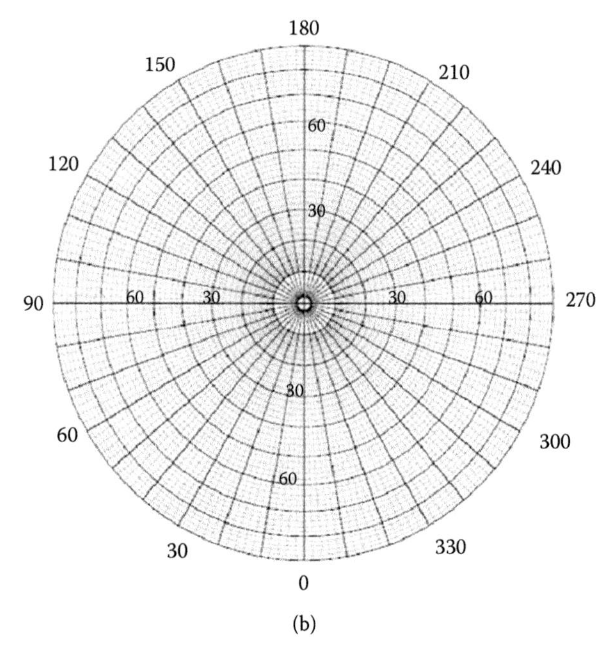

(b)

Figure A1.1 (Continued)

Index

For Product Safety Concerns and Information please contact our EU representative GPSR@taylorandfrancis.com
Taylor & Francis Verlag GmbH, Kaufingerstraße 24, 80331 München, Germany

www.ingramcontent.com/pod-product-compliance
Ingram Content Group UK Ltd.
Pitfield, Milton Keynes, MK11 3LW, UK
UKHW021119180425
457613UK00005B/147

* 9 7 8 0 4 1 5 8 0 9 2 3 8 *